Science in London

Istvan Hargittai · Magdolna Hargittai

Science in London

A Guide to Memorials

Springer

Istvan Hargittai
Budapest University of Technology
and Economics
Budapest, Hungary

Magdolna Hargittai
Budapest University of Technology
and Economics
Budapest, Hungary

ISBN 978-3-030-62335-7   ISBN 978-3-030-62333-3 (eBook)
https://doi.org/10.1007/978-3-030-62333-3

© The Editor(s) (if applicable) and The Author(s), under exclusive license to Springer Nature Switzerland AG 2021
This work is subject to copyright. All rights are solely and exclusively licensed by the Publisher, whether the whole or part of the material is concerned, specifically the rights of translation, reprinting, reuse of illustrations, recitation, broadcasting, reproduction on microfilms or in any other physical way, and transmission or information storage and retrieval, electronic adaptation, computer software, or by similar or dissimilar methodology now known or hereafter developed.
The use of general descriptive names, registered names, trademarks, service marks, etc. in this publication does not imply, even in the absence of a specific statement, that such names are exempt from the relevant protective laws and regulations and therefore free for general use.
The publisher, the authors, and the editors are safe to assume that the advice and information in this book are believed to be true and accurate at the date of publication. Neither the publisher nor the authors or the editors give a warranty, expressed or implied, with respect to the material contained herein or for any errors or omissions that may have been made. The publisher remains neutral with regard to jurisdictional claims in published maps and institutional affiliations.

Cover Illustration: Front cover art, "Newton" by Eduardo Paolozzi, 1995, after William Blake's "The Ancient of Days," 1805, British Library forecourt, 96 Euston Road, NW1 (photograph by the authors).

This Springer imprint is published by the registered company Springer Nature Switzerland AG.
The registered company address is: Gewerbestrasse 11, 6330 Cham, Switzerland

Foreword

London is rightly well known as a city of culture, business, politics, and the performing arts, but what is not so well appreciated is that London is also a city of science, in fact one of the great cities of science in the world. This is a book that celebrates the science and scientists over the ages who have had links to London through 'memorials', a wide range of artefacts and buildings including statues, portraits, plaques, buildings, museums, and scientific research institutions. These reflect the achievements and advances in science not only from those who came from or worked in the British Isles, but as befits a great global city, also many from around the rest of the world. The memorials span four centuries, ranging from Bacon and Newton in the seventeenth century through to Crick and Dirac in the twentieth.

The breadth of science reflected in these London memorials is quite extraordinary. The compass of science is drawn widely, covering the natural sciences, the applied sciences of medicine and engineering, as well as explorers who expanded humankind's experience of the natural world. The science and scientists covered also have great historical depth. It would not be an exaggeration to say that the birth of modern science took place in London with the establishment of the Royal Society of London, which was based on philosophical foundations of the seventeenth century, also closely linked to London. Collections relevant to science have been assembled in the city's great museums, such as the British Museum, the Natural History Museum, the British Library, and the Science Museum. These have engaged people in science from all walks of life and of all ages, including myself when as a child I visited these amazing institutions which introduced me to the wonders of the natural world. London not only has museums telling the stories of science but has world famous research institutions, many of great historical significance: the Royal Institution, Greenwich Observatory, Kew Gardens, and the great universities of University College, King's College, and Imperial College. Science in London is also to be found in the city's art galleries, a major example being the National Portrait Gallery.

I was brought up in London, and have worked there several different times during my life. Today, I still live and work in the city. Over the years, I have walked the streets of London always looking at the city through the prism of science. But despite my familiarity with London and the science of London, I have been astonished by what I have learnt from this book, what was revealed to me that I was not aware of previously. There is still much more for me to discover and for you to discover too. It is a book that will guide you to new places, to new memorials, to new people of science, and to see all of this through fresh eyes. It is comprehensive, erudite, and knowledgeable and a pleasure to read and to have as a guide for all those interested both in London and in science.

Nobel laureate in Physiology or Medicine 2001, Director of the Francis Crick Institute (London), former President of the Royal Society
London, UK

Paul Nurse

Preface

Science in London is our fourth volume in the series presenting the memorials to science and scientists in great cities, following similar volumes about Budapest, New York, and Moscow. We use "science" as an umbrella term covering explorations, medicine, and technology in addition to its traditional meaning of "natural" science. The leadership role of British science makes London an obvious choice for consideration. Cambridge, Oxford, Edinburgh, Manchester, and other cities have greatly contributed to science, but London is unique in two aspects. One is that it is the seat of the British monarchs whose interest facilitated and encouraged progress in science. The second is that London is the seat of great institutions, such as the British Museum, the Royal Society, and others. They have represented a focal point encouraging many of the great minds to congregate there. London, being the capital, has erected memorials to the greats of the nation regardless whether they lived in London or not. Scientists themselves recognized the importance of unhindered production and dissemination of knowledge; hence, they initiated great institutions in London. This was in contrast to many other great cities in the world where the authorities initiated the creation of institutions of science. Recognition of past performance led to an abundance of memorials ranging from conspicuous statues to the *blue plaques* (and plaques of other colors).

Royalty, military and political leaders, writers, artists, sportspersons, and other non-scientists have so many memorials in public places in London that those to scientists might appear rare. However, when we make an effort to look for them, they emerge in great abundance. The international outlook and imperial past have multiplied this presence. London's openness and the approach of integration rather than exclusion have further added to the success of cultivating science.

As we have issued similar warnings when discussing other cities, no realistic science history could be compiled on the basis of the memorials. Yet in the case of London, larger chunks of science history might be composed than elsewhere, due to the extraordinary number of blue plaques.

Chapter Overview

Explorers, scientists, medical professionals, and technologists/inventors can be distinguished easily, yet this grouping is also somewhat arbitrary because their areas often overlap. A number of scientists and inventors could as easily belong to one or another chapter. Some of the honorees represent different sides of related activities and could be discussed together had we followed a different scheme. Thus, for example, the physician John Snow and the civil engineer Sir Joseph Bazalgette both combatted the conditions leading to epidemics, but are discussed in two different chapters (with cross references, to be sure). Lord Kelvin presents a different kind of overlap between chapters. He in one person represents an equally great physicist and inventor. For these and similar reasons, consulting the index has added significance for mining information in this book.

Chapter 1: Introduction
The beginnings of modern science date back to the sixteenth and seventeenth centuries. This was so on the Continent, and Britain was no exception. Francis Bacon's philosophy of science prepared the ground for great science to come, which was assisted by influential institutions, such as the Royal Society. The Royal Society came to existence on the scientists' own initiative, but the reigns of some of the monarchs, such as Elizabeth I, George III, and Victoria and Albert, greatly facilitated general progress. The founding of the British Museum, also, started from private initiative. As its significance grew, it diversified and led to new institutions, such as the British Library, the Natural History Museum, and the Science Museum. A beautiful example of openness and assigning preeminence to competition rather than isolation was the Great Exhibition of 1851. Its consequences have reverberated ever since. This is expressed not only in the physically detectable structures of Albertopolis, but also in the 1851 scholarships that have fostered brilliant careers for so many gifted and diligent young people across the British Commonwealth.

Chapter 2: Explorers
The special geographic position of Britain and the curiosity and ambition of her people stimulated explorations around the globe from early on. Geography brought with it the progress in navigation and all sciences along with innovation related to its success. Memorials honor Prince Henry, the Navigator; Bartholomeu Dias; Sir Francis Drake; Sir Walter Raleigh; John Smith; Captain James Cook; Captain Matthew Flinders; Sir John Franklin; Frederick C. Selous; Robert Falcon Scott; David Livingstone; Ernest H. Shackleton; the Greenwich astronomers and inventors; and many others.

Chapter 3: Scientists—Physicists, Chemists, and Naturalists
Great minds toiled in the institutions of higher education in London; others were attracted by its open-minded atmosphere; yet others joined for social advancement. The desire to share knowledge was one of their motivations. Another was to contribute to solving practical problems. Science on the one hand, and industry, health care, and general economic considerations on the other, mutually strengthened each other. The Royal Society—the British national academy of sciences—is by itself a memorial. Its founders, among them Robert Boyle and Sir Christopher Wren (not only an architect, but also an astronomer), and its one-time employee— the first person ever making his living from science—Robert Hooke, and then, the greatest of all, Sir Isaac Newton, made it into the world's most coveted institution of science. The Royal Institution for "diffusing knowledge" has become what it is today through the activities of Sir Humphrey Davy, Michael Faraday, Mary Somerville, John Tyndall, Sir William Lawrence Bragg, Baron Porter, and others. Memorials to polymaths who excelled in more than one field, such as mathematics, physics, mechanics, and so on, include Pythagoras, Archimedes, Galileo, Newton, Leibnitz, Laplace, Goethe (the author, poet, and philosopher who was also a scientist), Thomas Young, Lord Kelvin, James Clerk Maxwell, and Baron Blackett. A few names must suffice from among the chemists that memorials honor: Joseph Priestley, Henry Cavendish, Antoine Lavoisier, John Dalton, Justus von Liebig, Dmitry I. Mendeleev, Sir William Ramsay, Sir William Henry Perkin, and Sir William Crookes. Memorials to naturalists and biologists honor some of the greatest scientists of all time: John Ray, Sir Joseph Banks, Carl Linnaeus, Georges Cuvier, Sir William Hooker, Charles Darwin, Alfred Russel Wallace, Thomas Henry Huxley, and many others. Contributors to DNA science also figure in our discussion, such as Francis Crick, James D. Watson, Rosalind Franklin, Maurice Wilkins, and their colleagues.

Chapter 4: Medicine Men and Women
The threat of epidemics, the desire to ease pain during surgeries, cures for common illnesses, increased longevity, sheer curiosity about the functioning of the human body, and lifting the general level of healthy living conditions were among the motivations that advanced health

care and prompted innovations in medicine. Guarding and elevating the quality of treatment led to developing institutions of various specialties of medicine. Prominent representatives commemorated by memorials include William Harvey, Edward Jenner, Hippocrates, Thomas Linacre, Sir Patrick Manson, and Sir Ronald Ross; pioneering women in medicine, such as Elizabeth Garrett Anderson, Elizabeth Blackwell, and Dame Louisa Aldrich-Blake; and a variety of medical professionals, among them Henry Gray, William Hunter, John Snow, Baron Lister, Sir Peter Medawar, Sir Henry Dale, Sir Alexander Fleming, John Langdon Down, James Parkinson, Robert Bentley Todd, Sigmund Freud, Galenus, John Hunter, and Sir Henry Solomon Wellcome.

Chapter 5: Innovators, Engineers, and Technologists
Moving information, goods, and people at ever-increasing speed and efficiency contributed to the innovations that brought about industrial revolutions. Memorials honor inventors in electricity and electronics; telecommunications; computation; mining and steel production; geology; chemical engineering; biotechnology; civil engineering; transportation; and innovations in chronometry as well as those who applied the new technologies for defense and invented holography. Some of the names of honorees are Leonardo, Benjamin Franklin, Sir Francis Ronalds, Lord Kelvin (again), Hertha Ayrton, Guglielmo Marconi, John Logie Baird, Charles Babbage, Countess Ada Lovelace, Alan Turing, William Smith, Henry Bessemer, Sir Charles Lyell, Alfred Nobel, Chaim Weizmann, John Rennie, James Walker, Sir Marc Isambard Brunel, Isambard Kingdom Brunel, Sir Joseph William Bazalgette, James Henry Greathead, James Watt, Richard Trevithick, Robert Stephenson, Sir George Cayley, Count von Rumford, Sir Henry Thomas Tizard, and Dennis Gabor.

The memorials manifest the inclusiveness of the British scientific community and also of British society with regards to immigrants. Furthermore, there appears hardly any political bias in recognizing achievement as there has been hardly any in letting all those gifted and willing to work thrive. There were a few exceptions; the time of the Restoration, in the 1660s, comes to mind only to underline the lack of political considerations over time. Another negative example is the persecution of some creative individuals on account of their sexual orientation. Alan Turing is a conspicuous example whose recognition in terms of memorials is also on the rise. A point here concerns women scientists who have been underrepresented in scientific life and in memorials, but, hopefully, this is changing. It is demonstrated by a number of examples in this book that science and the prerequisite education, whether formal or self-education, aided people of disadvantaged social status in crossing social boundaries and rising to high societal positions.

A large number of memorials are presented in the following pages. There are two additional venues relatively seldom mentioned though they represent an exceptional wealth of memorials to scientists. One is the National Portrait Gallery that does not charge admission to visit where the images of many of the scientists, explorers, medical people, and innovators are on display. The other is Westminster Abbey, serving as a mausoleum for the remains of many of the most distinguished Britons and displaying memorials to many whose remains rest elsewhere. It is possible to visit it for a fee. Both these venues eminently augment our collection.

A final note concerns the titles and nationalities that our description uses. We indicate "Sir" or "Dame" as they are part of the names, but rarely others unless some exceptional circumstance warrants it. Also, we do not distinguish the national origins among the British, viz., English, Scottish, Irish, or Welsh.

Also by the Authors

I. Hargittai, *Mosaic of a Scientific Life* (Springer Nature, 2020)

I. Hargittai, M. Hargittai, *Science in Moscow: Memorials of a Research Empire* (World Scientific, 2019)

B. Hargittai, Ed., *Culture and Art of Scientific Discoveries: A selection of István Hargittai's writings* (Springer Nature, 2019)

I. Hargittai, M. Hargittai, *New York Scientific: A Culture of Inquiry, Knowledge, and Learning* (Oxford University Press, 2017)

B. Hargittai, I. Hargittai, *Wisdom of the Martians of Science: In Their Own Words with Commentaries* (World Scientific, 2016)

I. Hargittai, M. Hargittai, *Budapest Scientific: A Guidebook* (Oxford University Press, 2015)

B. Hargittai, M. Hargittai, I. Hargittai, *Great Minds: Reflections of 111 Top Scientists* (Oxford University Press, 2014)

I. Hargittai, *Buried Glory: Portraits of Soviet Scientists* (Oxford University Press, 2013)

R.J. Gillespie, I. Hargittai, *The VSEPR Model of Molecular Geometry* (Dover, 2012)

I. Hargittai, *Drive and Curiosity: What Fuels the Passion for Science* (Prometheus, 2011)

I. Hargittai, *Judging Edward Teller: A Closer Look at One of the Most Influential Scientists of the Twentieth Century* (Prometheus, 2010)

M. Hargittai, I. Hargittai, *Visual Symmetry* (World Scientific, 2009)

M. Hargittai, I. Hargittai, *Symmetry through the Eyes of a Chemist* (3rd Edition, Springer, 2009, 2010)

I. Hargittai, *The DNA Doctor: Candid Conversations with James D. Watson* (World Scientific, 2007)

I. Hargittai, *The Martians of Science: Five Physicists Who Changed the Twentieth Century* (Oxford University Press, 2006, 2008)

I. Hargittai, *Our Lives: Encounters of a Scientist* (Akadémiai Kiadó, 2004)

I. Hargittai, *The Road to Stockholm: Nobel Prizes, Science, and Scientists* (Oxford University Press, 2002, 2003)

B. Hargittai, I. Hargittai, M. Hargittai, *Candid Science I–VI: Conversations with Famous Scientists* (Imperial College Press, 2000–2006)

Acknowledgments

The following provided a diverse kind of assistance in our work on this project, which we appreciate with gratitude: Philip Ball, Hannah Brown, Matt Brown, Polly Brown, Ralph R. Frerichs, Walter and Hannah Gratzer, Ella Griffiths, Michael Hunter, Sophie Johnson, Ian Jones-Healey, Géza Komoróczy, Rocio Lale-Montes, Shira Leibowitz Schmidt, Myer Leonard, Patrick Lewis, Alan L. Mackay, Robert Mackay, James Marshall, Salvador Moncada, László Papp, Holly Peel, Steve Roffey (Spudgun67), Henry Rzepa, John Smith, Laura Smith, and Tom Strom.

We have enjoyed the thoughtful and constructive assistance of Annamaria Szoke, Bob Weintraub, and Irwin Weintraub thoughout our entire work.

Our special thanks and appreciation are due to Timothy Bintrim for his creative contribution to shaping the entire text of our manuscript.

Executive Editor Charlotte Hollingworth of Springer encouraged us and assisted us on the way to creating *Science in London*.

We thank the Wellcome Foundation for the images we used, which are made available so generously in the Wellcome Collection.

Photographs for which no source is indicated were taken by the authors.

Contents

1 **Introduction**... 1
    Science and Technology in Unison....................................... 2
    Philosophical Foundation 3
    Royal Support 5
    Queen Victoria and Prince Albert 7
    British Museum.. 13
    British Library... 15
    Natural History Museum.. 16
    Science Museum... 17
    Burlington House 17
    Westminster Abbey... 20
    National Portrait Gallery.. 22
    Openness and Receptiveness 23

2 **Explorers**... 27
    Fifteenth through Seventeenth Centuries 32
    Australia 36
    Africa 38
    Polar Exploration 41
    Scientists as Explorers 46
    Greenwich.. 48

3 **Scientists**... 55
    Royal Society 57
    Royal Institution 68
    Physicists, Astronomers, and Mathematicians 79
    Chemists... 100
    DNA—Multidisciplinary Science....................................... 119
    Naturalists and Biologists... 123

4 **Medicine Men and Women**...139
    Royal College of Physicians... 144
    Royal Society of Medicine 148
    King's College 151
    St Thomas' Hospital 153
    Guy's Hospital 160
    London School of Hygiene & Tropical Medicine......................... 164
    Women in Medicine 170
    Nurses... 178
    Professions 179

5 **Innovators, Engineers, and Technologists**..............................217
    Electricity and Electronics 221

Telecommunications 222
Computing 230
Royal School of Mines (Imperial College) 235
Geology, Geodesy, Geography 238
Chemical Engineering (ICI—Imperial Chemical Industries) 240
Biotechnology 240
Civil Engineering 241
Steam Engine 251
Transportation 253
Clockmakers 261
Other Inventors 264

Bibliography 271

Index of Names 273

Index of Artists and Architects 279

Introduction

Sculpture of "Science" by Farmer and Brindley, 1868–1869, on the Holborn Viaduct, EC1

Science and Technology in Unison

"Science" in two representations. Left: one of the relief panels by John Hancock, 1864–1865, on the façade of the former National Provincial Bank, 15 Bishopsgate, EC2. Right: by John Birnie Philip on the façade of the Foreign and Commonwealth Office, Whitehall, WC2

In all three allegorical appearances above, a female figure represents Science. All have additional referents. The one on the Holborn Viaduct appears with the governors of the steam engine in her hands. There is a globe on a tripod at her left side encircled with the wire of an electric telegraph connected to a battery. There are references to astronomy and a crumpled paper with outlines of Pythagoras's theorem. In the panel on the façade of the bank building, the female figure appears with a scroll. On the left in this panel, a young, muscular workman is seated, while another, older man, resembling Sir James Watt,[1] seems to interpret. In a parallel scene on the right in this panel, an old man with a globe instructs two boys in navigation or geography. The allegorical figure in the Philip relief above has a quill in her right hand and a scroll in her left. Thus, in all three cases, Science is shown with practical applications, giving emphasis to the complementarity of science and technology.

---

[1] This reasonable suggestion is in Philip Ward-Jackson, *Public Sculpture of the City of London* (Public Sculpture of Britain Volume Seven, Liverpool: Liverpool University Press, 2003), p. 38.

Philosophical Foundation

Left: Statue of Sir Francis Bacon by Frederick William Pomeroy, 1920, in front of Gray's Inn, South Square, WC1. Right: Bas-relief panel by William Grinsell Nicholl, showing Sir Francis Bacon explaining his philosophy to his auditors, Oxford and Cambridge Club, 71–77 Pall Mall, SW1Y

Memorials of Sir Francis Bacon, left to right: by William Theed, Jr., in one of the ground-story niches on the right-hand side of Burlington House; by John Birnie Philip/Henry Hugh Armstead on the façade of the Foreign and Commonwealth Office; and by J. Daymound & Sons on the façade of the former building of the City of London School, Victoria Embankment, EC4

A Cambridge-educated philosopher, statesman, jurist, and author pioneering the scientific method, Sir Francis Bacon (1561–1626), laid the foundation of how science has been approached in Britain. His philosophy of science can be seen in the foundation of the Royal Society although by the time the Society came to be, he had been dead for decades. He created and advocated empiricism, insisting that knowledge comes from observation and experience. His scientific method, still called the Baconian method, was described in his book, *Novum Organum*, 1620. The title referred to Aristotle's *Organon* and was intended to replace it. Bacon emphasized experimentation instead of the Aristotelian approach of hypothesizing.

Recent examples demonstrated the timeliness of Bacon's teachings. William Harvey considered himself a conventional Aristotelian, yet he did away with the long-held views originating from Galen about how blood flowed in the body. Harvey discovered blood circulation and based his discovery on "ocular inspection."[2]

---

[2] See Adrian Tinniswood, *The Royal Society and the Invention of Modern Science* (New York: Basic Books, 2019), p. xvii.

Bacon applied inductive reasoning, starting with systematic observation and the collection of quality facts. From these facts, he proceeded to generalization. However, he stressed the importance of not going beyond the facts in such generalization. Any extension of such generalization necessitates additional observation and experience, that is, additional facts. Knowledge builds up in a stepwise fashion. The insistence of the Fellows of the Royal Society on conducting actual experiments, even during their meetings, reflected Bacon's teachings. His dedication to the empirical approach may have led to his demise: exploring how to preserve meat by freezing, during his experimentation, he contracted a fatal pneumonia.

Statues of John Locke by W. Theed, Jr., (left) and David Hume by M. Noble (right), both at Burlington House

John Locke (1632–1704), philosopher and physician, has been called the "Father of Liberalism" whose works influenced Thomas Jefferson when he wrote the *Declaration of Independence*. Locke studied at Westminster School in London and at the University of Oxford. We single out his contribution to epistemology. Among the pioneer empiricists, he followed Sir Francis Bacon. In Locke's *An Essay Concerning Human Understanding*, his main thesis was that we all are born with a blank mind. *Experience* gradually fills it up, through observation, the primary tool of acquiring knowledge and the principal tool of science. Any teaching should be testable without exception. In this, Locke anticipated the modern approach to scientific truth as Karl Popper stated that "*it must be possible for an empirical scientific system to be refuted by experience*"[3] (italics in the original). There is another John Locke statue by Richard Westmacott at the main building of University College London (UCL).

David Hume (1711–1776) was another philosopher known for empiricism in his enlightened system of philosophy, including epistemology. Having lost his father early, he could complete only part of his education at the University of Edinburgh. His book *A Treatise of Human Nature*, 1739–1740, was an attempt to create a complete science of man, which could serve as a roadmap to the understanding of all

---
[3] Karl Popper, *The Logic of Scientific Discovery* (London and New York: Routledge Classics, 2002; reprint of the 1959 first edition in English), p. 18.

the facets of human nature. His inductive reasoning was based on experience, that is, actual observation, from which conclusions and expectations could be drawn for situations not amenable to direct observation. Hume's activities were broad-based and most influential; suffice it to mention the example of Søren Kierkegaard's rational theology and today's cognitive science in which Hume's long-lasting impact is discernable.

Royal Support

Left: Portrait of Queen Elizabeth I at Wanstead Palace, Mark Gheeraedts (Wellcome Collection). Right: Bas-relief by John Birnie Philip/Henry Hugh Armstead on the façade of the Foreign and Commonwealth Office

William Shakespeare's statue by Giovanni Fontana, after Peter Scheemakers, 1974, on Leicester Square, WC2. The enlarged detail is a quote from the Clown in Shakespeare's *Twelfth Night*

Elizabeth I (1533–1603) reigned as Queen of England and Ireland, 1558–1603. Despite much turmoil, her Elizabethan era has been regarded as prosperous. One of her statues stands in a niche of the façade of the St Dunstan-in-the-West Church (not shown here). This was not the original location of the statue as the present building of the church dates back to the 1830s. The statue was carved in the Queen's lifetime and is the oldest outdoor statue in London. There was a remarkable cultural life in London under her reign, and William Shakespeare (1564–1616) was a major contributor to it. The original of his statue on Leicester Square stands in Poets' Corner of Westminster Abbey. There, the scroll has a longer quotation from *Twelfth Night*, whereas Fontana's version bears an epigrammatic, pointed message. It calls for escaping ignorance by acquiring knowledge, lifting oneself out of darkness.

Under the rule of Elizabeth, some European instrument makers received special privileges in her realm. Because they set up shop in London, many skilled English craftsmen learned their trade from them. Innovation in instruments has always been a driving force in scientific progress and technological advancement. It also greatly facilitated exploration (Chap. 2). The Elizabethan provision for instrument makers was one of the roots of the emergence of enhanced interest in science. Her reign and the next decades provided the conditions for Francis Bacon's philosophy about the scientific method; the establishment of the Royal Society following the Restoration; and, eventually, Isaac Newton's milestone discoveries.

Left: William Gilbert demonstrating a magnet before Queen Elizabeth, 1598. Oil painting by Ernest Board. Right: John Dee performing an experiment before the Queen. Oil painting by Henry Gillard Glindoni (both, Wellcome Collection)

The two paintings above illustrate Queen Elizabeth I's interest in science and innovation. Such events at the Court provided encouragement for such activities. William Gilbert (1544–1603) trained as a physician. Toward the end of his life, he was appointed physician to the Queen, but his most important work was in the study of magnetic bodies and the nature of electricity. He is often referred to as the father of the study of electricity and electrical engineering. Also a gifted astronomer, he agreed with Copernicus that the Earth rotates on its axis. Gilbert was not aware of gravitation so he thought that magnetism held the planets in their orbits. John Dee (1527–1608) was a Cambridge-educated scientist who augmented his training on the continent. An adviser to

Queen Elizabeth I in science and medicine, he built his own laboratory, conducted experiments, and established his own library. He assisted exploration and provided instructions in navigation and cartography as well as navigational instruments to the voyages of exploration sponsored by Elizabeth.

The only deed by King Charles II (1630–1685) in connection with science was his issuing a Royal Charter for the Royal Society, but this was, of course, a prerequisite for the Society to become Royal. He reigned from 1660 to 1685, and his charter of the Royal Society was signed early in his reign. After his father Charles I was executed in 1649, Charles II lived in exile until, upon Oliver Cromwell's death in 1658, he was called back to the throne in 1660. Not all his biographies mention his role in establishing the Royal Society, but for our story, this was significant. His bust stands in the lobby of the Royal Society (Chap. 3).

The reign of King George III (1738–1820) lasted from 1760 close to 60 years, though during its last decade, he was placed under regency on account of his mental illness. He had a unique education for a future monarch: his studies included French and Latin, in addition to English and German; chemistry, physics, astronomy, and mathematics; and history, geography, commerce, agriculture, law, and religion, and he also was accomplished in music, dancing, fencing, and riding. He developed the King's Library and made it available to scholars. It became a component of the British Library, which exhibits his bust (see below). He collected mathematical and scientific instruments, which have been deposited in today's Science Museum. Not only did he found his own observatory but also supported the construction of William Herschel's 40-foot telescope, which was then the world's largest. Agriculture, science, and technology made great advances during his reign. We have emphasized the connection of George III to science; however, the history of his reign is better known for a number of tumultuous events; suffice it to mention the American Revolution and War of Independence, the Napoleonic wars, and the forced transportation of about 1.8 million enslaved persons out of Africa to British colonies.

Queen Victoria and Prince Albert

Statue of Queen Victoria unveiled in 2007 on the occasion of the centenary of Imperial College London, Exhibition Road, SW7

Queen Victoria (1819–1901) ascended to the throne in 1837, and her reign was a golden period for British exploration, science, and technology. Prince Albert, the Prince Consort (1819–1861), married his cousin in 1840, after she had proposed to him in 1839. Prince Albert had a difficult road to influence and popularity, but slowly and gradually, he succeeded. He was interested in modernizing the royal household as well as the country.

Michael Faraday lecturing at the Royal Institution, before Prince Albert and his two sons, Edward, the Prince of Wales, and Prince Alfred. From a sketch by Alexander Blaikley, 1856 (Wellcome Collection)

Prince Albert was elected, though not without opposition, Chancellor of the University of Cambridge. In this role, he campaigned for educational reform, the introduction of instruction in modern history and the natural sciences. He referred to the example of his alma mater, the University of Bonn. Prince Albert was not only interested in science; he was active in its institutions, among them the British Association for the Advancement of Science. For some time, he served as the president of the Association. The image above shows Prince Albert and his two sons, Edward and Alfred, attending Michael Faraday's lecture at the Royal Institution. Temperamentally interested in science, he took his two sons, but not his daughters, to hear Faraday.

Prince Albert realized that the British manufacturing sector could benefit from more competition and made this one of the driving ideas behind the Great Exhibition of 1851. Although all kinds of misgivings preceded the event, among them that it would cost too much and would result in considerable loss, it became a resounding success. Prince Albert found a dedicated and able assistant in making this great event happen in the person of Henry Cole.

Left: The Great Exhibition of 1851 in the Crystal Palace, Hyde Park: the transept looking north inside and outside. Steel engraving by W. Lacey after J. E. Mayall, 1851 (Wellcome Collection). Right: View from the Knightsbridge Road of the Crystal Palace (Wikimedia)

Sir Henry Cole (Wikimedia) and his plaque, 33 Thurloe Street, SW7

Sir Henry Cole (1808–1882) left school at 15, but thanks to his skill in writing and drawing, was employed by the records office and rose gradually in the civil service. He initiated and edited the *Post Circular* and became involved with postal stamp design. He was an inventor, did prize-winning industrial design, and wrote books for children, all under the pseudonym Felix Summerly. He popularized sending greeting cards and introduced the first Christmas card in 1843. Later, he was the first director of the Victoria and Albert Museum and lived nearby, opposite the Museum.

Cole participated in organizing exhibitions and noticed the need to introduce international participation in them. He managed the activities of the Royal Commission for the Exhibition of 1851 established by Queen Victoria under the presidency of Prince Albert. A most fruitful cooperation developed between Prince Albert and Cole. When the Exhibition in the Crystal Palace (a temporary structure) in Hyde Park produced a huge cash surplus, Cole was instrumental in the decision to use it for improving science and art education.

The surplus was used eventually to purchase land in South Kensington to establish the Natural History Museum, Science Museum, Imperial College, the Royal Albert Hall, and the Victoria and Albert Museum. A small portion of the Exhibition proceeds was used for scholarships and has been ever since aiding dedicated young people to pursue higher education. The Science Research Scholarship, established in 1891, benefitted a number of future internationally renowned scientists and afforded a number of future Nobel laureates extended opportunities for study in the best schools under most favorable conditions.[4]

Bas-reliefs on the façade of 219 Oxford Street at the corner of Oxford Street and Hill Place, W1D (on the façade, they are from top to bottom). They depict, among others, the Royal Festival Hall, the logo of the Festival of Britain, the Dome of Discovery, and the Skylon Tower

Although the 1951 Festival of Britain was in some way a reverberation of the 1851 Great Exhibition, it was on a more modest scale; it was British rather than international; and it was in a country that had suffered the losses and destruction of World War II. However, it was also a show of defiance so soon after the war and in celebration of British science, technology, and art. It took place mostly on the South Bank of the Thames. The bas-reliefs on the façade of 219 Oxford Street refer to some of the technological achievements displayed by the Festival. The Royal Festival Hall was there to stay, altered subsequently, whereas the Dome of Discovery was a temporary exhibition venue. The cigar-shaped Skylon aluminum-clad steel tower was another temporary construction. It was based on the so-called *tensegrity* concept—integrity structure under tension—kept together by cables. The Skylon was dismantled soon after the Festival; today the Museum of London (150 London Wall, EC2Y) displays its model.

---

[4]From among our acquaintances alone we can name several: Sydney Brenner and Aaron Klug of South Africa and Mark Oliphant, John W. Cornforth, and Rita Harradence of Australia, all exceptional scientists, benefited from 1851 Scholarships.

Memorials of Prince Albert; from left to right: by Joseph Durham, 1863, in front of the Royal Albert Hall, in commemoration of the Great Exhibition 1851; by Charles Bacon, 1874, at Holborn Circus; and by John Henry Foley and Sir Thomas Brock, 1876, in the center of the Albert Memorial designed by Sir George Gilbert Scott, 1872, Kensington Gardens, SW7, close to Kensington Gore and north of the Royal Albert Hall

Of the many Albert memorials, we single out three for special reasons. The one at the Royal Albert Hall was meant originally just to honor and commemorate the Great Exhibition 1851. In time, it was decided to top it with a Prince Albert statue. The memorial at Holborn Circus depicts an equestrian statue of Prince Albert. The inscription of the relief panel on the south side of the pedestal is EXHIBITION OF ALL NATIONS, 1851. An allegorical Britannia is presiding with a lion and the representatives of the exhibiting nations in their national costumes.

The most conspicuous is the Albert Memorial in Kensington Gardens. We present the sculptures of four continents that stand in four corners of the memorial in Chap. 2. The central structure has four additional sculpture groups symbolizing Agriculture (southwest), Engineering (northwest, Chap. 5), Commerce (northeast), and Manufacture (southeast). The central structure is decorated by two sets of sculptures. At the upper level in four niches appear Rhetoric, Medicine, Philosophy, and Physiology. At the lower level on four pillars: Geometry, Geology, Astronomy, and Chemistry. Those that are relevant to the materials in subsequent chapters are presented there.

The most exciting component of the Albert Memorial is the so-called Frieze of Parnassus, constituting 169 mostly full-size sculptures, grouped according to the four sides of the Memorial: musicians and poets (south); sculptors (west); architects (north); and painters (east). The list of the 169 greats is impressive and admirable. It is a sort of international Panthéon[5] *for the arts*. Knowing, however, Prince Albert's interest in science, engineering, and inventions, their omission is regrettable. The name *Parnassus* usually refers to the home of poetry. However, extending it to learning and knowledge, further areas of human endeavor could have been added. It would be an interesting game to compile Frieze of Parnassus selecting explorers, scientists, people of medicine, and innovators/technologists.

At closer examination, even the existing Frieze contains some individuals who may be considered for this book. Pythagoras is shown among the poets, and we mention him for his geometry in Chap. 3. Johann Wolfgang von Goethe appears among the poets, most properly, but he was also a significant scientist and we mention him also in Chap. 3. Albrecht Dürer figures among the painters in the Frieze, but his mathematics was also noteworthy. Leonardo da Vinci is among the painters, while in our discussion he appears in Chap. 5. Many of the greats of the past were greats in more than one field; sometimes they even had to be because, for example, an architect had to be versed in several areas of his trade that would be rather alien to today's specialized experts. Sir Christopher Wren, who appears among the architects, was more than a multifaceted architect: he was also an anatomist, astronomer, mathematician, and scientist, and a founding member of the Royal Society. He appears in Chap. 3. There are others among the 169 whose noteworthy activities extended well beyond the arts.

---

[5]During the French Revolution, the Paris Panthéon (a temple dedicated to all the gods) was transformed from a church into a secular mausoleum for the remains of distinguished French citizens.

Temple Bar Memorial on Fleet Street at Bell Yard, WC2, in front of the Royal Courts of Justice, 1880, with Queen Victoria's statue flanked by the vertical sets of symbols of SCIENCE and ART

That science and technology were success stories in the Victorian era is manifested by their frequent presence in the memorials of royalty. The Temple Bar Memorial is an example. Its location is where the original Temple Bar, designed by Sir Christopher Wren, used to stand. This Memorial, with a dragon at the top, was designed by Horace Jones. The statues of Queen Victoria and the Prince of Wales—later Edward VII—were created by Joseph Boehm. There is direct relevance of this memorial to our discussion in that sets of reliefs SCIENCE and ART decorate both niches of the statues, thus giving elevated importance to these two domains of human activity. The SCIENCE column displays symbols of physics, geography, astronomy, medicine, engineering, chemistry, and mechanics.

British Museum

Memorials of Sir Hans Sloane, from left to right: his statue by Simon Smith, 2007, Duke of York Square, SW3; his plaque, 4 Bloomsbury Place, WC1A; and his bust by Michael Rysbrack at the British Museum

Sir Hans Sloane (1660–1753) was a polymath and a collector. On account of his bequest of his exceptionally rich collection to the nation, he is considered the founder of the British Museum, the British Library, and the Natural History Museum. From early youth he collected objects of natural history, manuscripts, and books. He studied medicine in London and became an MD. He was elected Fellow of the Royal Society when he was 24. He participated in expeditions to Jamaica and elsewhere. He is credited with first introducing milk chocolate as a drink to England. Made wealthy from the expeditions and by marriage, he was created baronet, the first medical practitioner receiving such recognition. He was President of the Royal College of Physicians and, as Isaac Newton's successor, President of the Royal Society. Recently, his bust was removed from its prominent place at the "Enlightenment Gallery" to a separate display about him as "collector and slave owner."

Main building and entrance portico, British Museum, Great Russell Street, WC1

Views of one of the exhibition halls, "the Enlightenment Gallery"

The British Museum was founded as a "universal museum" by Royal Assent of King George II and the Act of Parliament in 1753. It was the world's first national museum. The present main block and façade at Great Russell Street was designed by the architect Sir Robert Smirke and constructed 1823–1846. Today, the British Library and the Natural History Museum that both used to be parts of the British Museum are independent entities, not only by location but also administratively. The three have a total of 233 million objects, of which 13 million are at the British Museum and 70 and 150 million at the Natural History Museum and the British Library, respectively. Still, the British Museum has retained its universality and represents both modern and ancient civilizations.

Museum Tavern, opposite the British Museum, has its sign depicting Sir Hans Sloane's portrait. Its customers included such notables as Karl Marx, the author of *Das Kapital*; Sir Arthur Conan Doyle, the creator of the Sherlock Holmes detective stories; and the playwright J. B. Priestley.

British Library

From left to right: Busts of Sir Robert Cotton by Louis Roubiliac; Sir Hans Sloane by Michael Rysbrack; King George III by Peter Turnerelli; Sir Joseph Banks by Anne Seymour Damer; and Thomas Grenville by G. B. Comolli, at the British Library

The initial collection that formed what is today the British Library came from books and manuscripts of Sir Robert Cotton (1571–1631), Sir Hans Sloane (1660–1753), Robert Harley (Earl Mortimer, 1661–1724), King George II (1683–1760), and King George III (1738–1820). George III's collection was called the King's Library. For Sir Joseph Banks, see Chap. 3 (Section: Naturalists and Biologists). Thomas Grenville (1755–1846) was a politician and bibliophile. He bequeathed his 20,000-volume library to the British Museum.

Today, the British Library, the United Kingdom's national library, has well over 170 million items, books, and artifacts. It used to be part of the British Museum and became independent in 1973. The present location opened in 1998 on the north side of Euston Road, between Euston Railway Station and St Pancras Railway Station. Its emblematic Newton sculpture, after William Blake by Eduardo Paolozzi, stands in its piazza (see Chap. 3 and is on the front cover of this book).

Natural History Museum

The Natural History Museum, as viewed from Cromwell Road (left) and Exhibition Road (right), SW7

Views of the Central Hall and the mineral collection, Natural History Museum

The origin of the Natural History Museum dates back to the scientific collection of Sir Hans Sloane. The Museum was founded in 1754, but the construction of its present venue according to the design of the architect Alfred Waterhouse began in 1873 and was completed in 1881. The Museum has specimens related to life and earth science. It consists of five collections: botany, entomology, mineralogy, paleontology, and zoology. For its magnificent design, it is often referred to as the "Cathedral of Nature." It gained administrative independence from the British Museum in 1992. Its magnificent Central Hall displays Charles Darwin's statue (Chap. 3) and a blue whale skeleton hanging from the ceiling.

Science Museum

Science Museum, Exhibition Road, SW7, and an interior view of the Museum

The Science Museum was founded in 1857. Today it has 3.3 million visitors annually. Over the years, on our numerous visits, we always found it crowded. The Museum displays many original prototypes of milestone developments of science and technology. There are Stephenson's oldest surviving steam locomotive, Watson and Crick's reconstructed double helix model of DNA, Babbage's computing engine, a model of the Russian Shukhov's hyperbolic tower, and many more. Some images from the Science Museum will figure in subsequent chapters.

renovations over time. The British Government bought it in 1864, and for a long time the Royal Society and other learned societies were headquartered there. Today it is home to the Geological Society, the Linnean Society, the Royal Astronomical Society, and the Royal Society of Chemistry, in addition to the Royal Academy of Arts and the Society of Antiquaries of London.

There are 22 statues on the Burlington Gardens side and 8 more on the court side. Those appearing in further chapters are in bold in the list below.

Burlington House

The Burlington House was built in the 1660s as a private mansion, and at some time it belonged to the Earls of Burlington, hence its name. It has gone through a series of

The left, central, and right portions of Burlington House, 6 Burlington Gardens, W1

The statues on the façade on Burlington Gardens are, when facing Burlington House, on the left, "Illustrious Foreigners," on the balustrade (by E. W. Wyon):

Galilei, Galileo ("Galileo," 1564–1642), Italian astronomer, natural philosopher

Goethe, Johann Wolfgang von (1749–1832), German poet, author, and scientist

Laplace, Pierre Simon ("La Place," 1749–1827), French mathematician and astronomer. In the ground-story niches (by Patrick McDowell)

Leibnitz, Gottfried Wilhelm (1646–1716), German philosopher and mathematician

Cuvier, Georges (1769–1832), French zoologist and anatomist

Linnaeus, Carolus (Carl von Linné, 1707–1778), Swedish naturalist

Statues of Aristotle by James Sherwood and J. S. Westmacott and Plato by William Frederick Woodington at Burlington House

On the central balustrade, representatives of ancient culture (Galen, Cicero, and Aristotle by James Sherwood and J. S. Westmacott; Plato, Archimedes, and Justinian by William Frederick Woodington):

Galen (Galenus), Claudius (c.130–c.201), Greek physician
Cicero, Marcus Tullius (106–43 BCE), Roman statesman and author
Aristotle (384–322 BCE), Greek philosopher and scientist
Plato (c.428–c.348 BCE), Greek philosopher
Archimedes (c.287–212 BCE), Greek scientist
Justinian (c. 482–565), Roman emperor and legislator

In the center, above the portico (by Joseph Durham):

Newton, Sir Isaac (1642–1727), scientist and mathematician

Bentham, Jeremy (1748–1832), philosopher and social reformer
Milton, John (1608–1674), poet
Harvey, William (1578–1657), physician

On the right, "English worthies," on the balustrade (by M. Noble):

Hunter, William (1718–1783), physician
Hume, David (1711–1776), philosopher
Davy, Sir Humphry (1778–1829), chemist

In the ground-story niches (by W. Theed, Jr.):

Smith, Adam (1723–1790), economist
Locke, John (1632–1704), philosopher
Bacon, Sir Francis (1561–1626), philosopher

Court-side façade of Burlington House with seven statues in second-story niches

The statues on the façade from the court side:

Leonardo (Leonardo da Vinci (1452–1519), Italian artist, engineer, and inventor
Flaxman, John (1755–1826), sculptor
Raphael (Raffaello Santi, 1483–1520), Italian painter
Michelangelo (inscribed as Michail Angelo, 1475–1564), Italian artist

Titian (Tiziano Vecellio, c.1488–1576), Venetian painter
Reynolds, Sir Joshua (1723–1792), painter, whose bronze statue (by Alfred Drury) stands in addition in the courtyard
Wren, Sir Christopher (1632–1723), architect, astronomer, and anatomist

Westminster Abbey

Westminster Abbey, Dean's Yard, SW1, and the interior of its Henry VII's Chapel, engraving by J. Jackson after W. F. Smallwood, 1843 (Wellcome Collection)

Westminster Abbey is one of the best-known churches in the world. Built in 960, it has been the traditional venue for the coronation and the burial of English, later British, monarchs. It has also been a burial place for many of the greatest contributors to British political and cultural life, the military, and science and technology. It has been often referred to as the British Valhalla, Valhalla being the burial place of gods in Nordic mythology. It may be similarly apt to compare it to the *Panthéon* in Paris and the *Novodevichy Cemetery* in Moscow.[6] Many greats whose remains rest elsewhere are commemorated here as well. We note with regret that taking photographs is not allowed in Westminster Abbey. We present here three images from the Abbey courtesy of the Wellcome Collection.

The Henry VII's Chapel whose interior is shown above, contains, among others, the Royal Air Force (RAF) Chapel. Among its many memorials, there is one to Sir Frank Whittle (1907–1996), pioneer of the jet engine. His Whittle gas turbine is mentioned in Chap. 5 among the British technological advances introduced to the Americans during WWII in the framework of the US-British cooperation. Whittle is buried at Cranwell, Lincolnshire.

---

[6] I. Hargittai and M. Hargittai, *Science in Moscow: Memorials of a Research Empire* (Singapore: World Scientific, 2019), Chapter 7, pp. 254–303.

Monuments of Sir Isaac Newton, 1731 (left), and the first Earl of Stanhope, 1733 (right), in Westminster Abbey. Colored aquatint by A. Pugin and T. Rowlandson(?) after W. Kent and M. Rysbrack (Wellcome Collection)

There is a conspicuous twin memorial in the center of the Nave designed by W. Kent and executed by J. M. Rysbrack. The one to the north of the entrance into the Choir is the memorial and burial place of Sir Isaac Newton (see Chap. 3). The gravestone carries a Latin inscription whose English translation is "Here lies that which was mortal of Isaac Newton." A cluster of graves and memorials of other great scientists is in close vicinity of the Newton memorial, viz., Paul Dirac, J. J. Thomson, Ernest Rutherford, Lord Kelvin, George Green (mathematical physicist of Green-functions fame), Michael Faraday, and James Clerk Maxwell. There are many more somewhat farther away. The memorial to the south of the entrance into the Choir is of the first Earl of Stanhope (1673–1721), military commander and statesman.

Funeral of Charles Darwin in Westminster Abbey (Wellcome Collection)

The funeral of Charles Darwin (Chap. 3) took place on April 26, 1882, at Westminster Abbey. Sir Joseph Hooker, Thomas H. Huxley, and Alfred Russel Wallace (all three figures in Chap. 3) were among the pallbearers.

National Portrait Gallery

When the National Portrait Gallery opened in 1856, it was the world's first portrait gallery. Since 1896, it has occupied its present location at St Martin's Place, WC2H. Only a fraction of its 300,000 paintings are on display at any given time. Scientists, inventors, and others who constitute the subject matter of this book are well represented.

Openness and Receptiveness

Plaques of Sir William Beveridge, 27 Bedford Gardens, W8, and of A. V. Hill, 16 Bishopswood Road, N6 (Spudgun67)

Beautiful examples of openness and receptiveness of British society were manifested at the time of anti-Semitic persecution of scientists and other Jews in Germany and German-occupied territories in the 1930s. When Germany dismissed Jewish scientists upon the Nazi takeover in 1933, British academia formed an organization of assistance, the Academic Assistance Council (AAC) at the urging of the economist William Henry Beveridge (later, Lord Beveridge, 1879–1963). He has been referred to as "the architect of the Welfare State," which nickname acknowledges the significance of his political activities. The Royal Society provided office space for the AAC, and prominent scientists joined it. The Nobel laureate Lord Rutherford was its president. The pavement mosaic titled "Curiosity," shown below, honors Rutherford with a profile suitable for a king. The background scene though might represent nuclear physics, which was the domain of Rutherford's major discoveries. Another Nobel laureate, A. V. Hill, was the vice-president of the AAC. Three Nobel laureates, Sir Frederick Gowland Hopkins, Lord Rayleigh, and Sir William Henry Bragg, along with J. B. Haldane, were council members.

A. (Archibald) V. (Vivian) Hill (1886–1977) graduated from Trinity College, Cambridge, where he excelled in mathematics and physiology—a rare combination. Following service in WWI and appointments at Cambridge and Manchester, he was Professor of Physiology at UCL until his retirement in 1951. He made fundamental contributions to the emerging science of biophysics and to operational research. His discoveries about the mechanism of heat production and mechanical work in muscles earned him a share of the 1922 Nobel Prize in Physiology or Medicine, jointly, with Otto Fritz Meyerhof. In 1935, he served on the committee charged with developing radar. During WWII, he worked for mobilizing Allied scientists for the war effort. After the war, he continued his research in biophysics and his service in the scientific community at home and internationally.

Pavement mosaic "Curiosity" by Boris Anrep in the Portico of the National Gallery, Trafalgar Square, WC2N. Portrait of Albert Einstein by Jacob Epstein, 1933, at the Victoria and Albert Museum, Cromwell Road, SW7

On October 3, 1933, a large meeting at the Royal Albert Hall organized by AAC and other aid groups was attended by 10,000 people. Albert Einstein gave a rousing anti-Nazi speech in defense of individual liberty. He told his audience, "You have shown that you, and the British people as a whole, have remained faithful to the tradition of tolerance and justice which your country has proudly upheld for centuries."[7]

It was during Einstein's brief sojourn in Britain that the American-British sculptor Jacob Epstein created his famous bust of the physicist. The artist noted, "his glance contained a mixture of the humane, the humorous, and the profound."[8] Among the refugee scientists granted asylum, 16 became Nobel laureates, 18 were knighted, and over a hundred were elected Fellows of the British Academy or the Royal Society.[9]

---

[7] Quoted in Abraham Pais, *Einstein lived here* (Oxford: Clarendon Press, 1994), p. 195.

[8] From the tag accompanying the bust at the Victoria and Albert Museum in a temporary exhibition in 2019, where the bust was on loan from Tate.

[9] The Academic Assistance Council has continued its activities ever since. Following name changes, its current name is Council for At-Risk Academics.

Memorials of the *Kindertransport* inside the Liverpool Street Station (detail) and at the Westbahnhof of Vienna, both by Flor Kent

The Nazis imposed a nationwide, large-scale anti-Jewish pogrom, the *Kristallnacht*, on November 9–10, 1938, in Germany, which by then had annexed Austria. Immediately afterward, 10,000 Jewish children were saved by the British people, bringing them from Germany and German-occupied territories to the United Kingdom. Called the *Kindertransport*, this timely action has been commemorated not only in London where the children were received but also in Berlin, Vienna, and other occupied cities, from whence these children were saved.[10] In London, the main receiving venue was the Liverpool Street Station, which has poignant memorials to the *Kindertransport* both inside and in front of the Station. A number of world-renowned future scientists were among those children.

---

[10] Two of the authors' future friends were among those children. Alfred Bader (1924–2018), in part of Hungarian origin, was a Canadian-American chemist, art collector, and philanthropist. He grew up in Vienna. He accompanied one of the authors (IH) on a visit to the school in Vienna, which was the *lager* (slave labor camp) in 1944–1945, where IH and his family were incarcerated. The other, the American physicist Arno Penzias (1933–), received the Nobel Prize in Physics in 1978 for the discovery of the background microwave radiation. He wrote the Preface for one of the authors' books.

Statue of Raoul Wallenberg by Philip Jackson at Great Cumberland Place, W1, between Berkley Street and Seymour Street, and the last known photograph of Wallenberg (courtesy of the late Lars Ernster)

Raoul Wallenberg (1912–1947?) saved thousands of Jews in Budapest during the second half of 1944. He joined the Swedish Embassy in July 1944 as a low-ranking diplomat. By then the Hungarian Jews from outside of Budapest had been deported and largely exterminated, but there still remained a considerable Jewish population, about 200,000 people, in the capital city. Sweden's behavior during WWII was not exemplary, and there were some in Sweden who wanted their country to provide asylum to those persecuted by the Nazis. Yet Wallenberg's heroism went far beyond any definition of duty. He spent the last days of the Battle of Budapest in a cellar, which he left on January 17, 1945, to reach out to the approaching Soviets. After being detained by Soviet counterintelligence, he and his driver perished somewhere in the Soviet Union. Today, there are memorials honoring Wallenberg all over the world. Among the many Jews Wallenberg saved, some became well-known scientists.[11]

---

[11] A few examples: Lars Ernster (1920–1998) was excluded from higher education in Hungary. In 1944 he was saved by Wallenberg from the hands of the Hungarian Nazis. Ernster and his wife moved to Sweden in 1946. He studied at the University of Stockholm, earned his PhD degree, and became a professor of biochemistry at his alma mater. He was a member of the Royal Swedish Academy of Sciences; biochemistry and cell biology were his principal research areas. He served on the Nobel Prize Committee for Chemistry between 1977 and 1988 and as a member of the Board of Trustees of the Nobel Foundation in 1990–1991. Gabor Somorjai (1935–) was also saved by Wallenberg. Somorjai immigrated to the United States following the suppressed Hungarian revolution in 1956. He completed his studies at the University of California at Berkeley, where he spent much of his professional life. He has been a world leader in surface science and catalysis. He received the Wolf Prize in Chemistry and the US National Medal of Science. Wallenberg rescued Francis Dov Körösy (1906–1997) just as the Hungarian Nazis were about to shoot him into the Danube. Relocating to Israel in 1957, Körösy became an internationally renowned physical chemist and served as the director of the Chemical Laboratory at the Negev Institutes for Desert Research.

Explorers 2

Statue of Captain James Cook by Thomas Brock, 1914, on the Mall, close to the Admiralty

Much of the allegorical expression of exploration appears as navigation, in keeping with the saying, *Navigare necesse est*… "To sail is necessary."[1] In British art, exploration is most often represented as marine navigation rather than land, aeronautic, or space navigation. A number of sculptures are dedicated to navigation in London. Atop the Foreign and Commonwealth Office, a female allegorical figure holds a rudder. On the Admiralty Arch, its principal female figure symbolizing Navigation holds a sextant in her hands. There is another similar figure at Trinity Square.

Four allegorical statues flank the statue of Queen Victoria by J. Birnie Philip at the top of the Foreign and Commonwealth Office, Whitehall, WC1: "Legislation" by Henry Hugh Armstead; "Wisdom" by J. Birnie Philip, on the left; "Justice" by J. Birnie Philip; and "Navigation" by Henry Hugh Armstead, on the right

"Navigation" by Sir Thomas Brock, 1911, at the Admiralty Arch, SW1, and a group of young explorers at the sculpture

---

[1]The full Latin quotation is *Navigare necesse est, vivere non est necesse*—"To sail is necessary, to live is not necessary." These words are attributed to the Roman statesman Pompey the Great who thus inspired the sailors to sail even in stormy weather to bring food from Africa to Rome.

"Navigation" by Henry Hugh Armstead, at the top of the *Foreign* and Commonwealth Office, as a woman with a rudder (left), and by Charles Doman at the former Port of London Authority, Trinity Square, EC3, as a woman grasping a ship's great wheel with her foot resting on a globe

In a broad sense, all scientists are explorers: finding new worlds on the Earth and in the heavens, finding cures for diseases, creating machines to make life easier, working in their laboratories, or just thinking in their offices. In this chapter, we introduce explorers in the traditional meaning of the word: those who visited faraway lands to explore them. However, some of the scientists mentioned in this chapter did more than just that, so it is necessary that some of the names will appear in more than one chapter.

Continents at the Albert Memorial, from left to right: "Europe" by Patrick MacDowell; "Africa" by William Theed; "Asia" by John Henry Foley; and "America" by John Bell, all between 1865 and 1871

British exploration of the continents is commemorated in at least two places in London. Facing the Albert Memorial from the south, the "Europe" sculpture group is in front on the left (southwest corner), representing Britannia, France, Germany, and Italy. "Africa" is in front on the right (northeast corner) representing an Egyptian woman on a camel, a servant boy with his hand resting on a sphinx, a seated Arab man, and behind him, a woman and an African man. "Asia" is in the back on the right (southeast corner), representing an Indian elephant, flanked by an Indian soldier, a Persian poet, a Chinese porcelain maker, and a Turkish merchant. "America" is in the back on the left (northwest corner), representing the United States, Canada, Mexico, and South America.

Foreign and Commonwealth Office, Whitehall, SW1

Bas-reliefs by Henry Hugh Armstead in the spandrels on the first story on the façade of the Foreign and Commonwealth Office, Whitehall, SW1, represent "Europe," "Africa," "Asia" (top row, from left to right), "America," and "Australasia" (bottom, left and right)

Our other example of the continents is at Whitehall, all as female allegorical figures. "Europe" is represented with a ship and a horse; "Africa" with a child, a banana tree, and a hippopotamus; "Asia" with an elephant; "Australasia" with sheep and kangaroo; and "America" with feathers and a spear.

Fifteenth through Seventeenth Centuries

Prince Henry the Navigator, Infante Dom Henrique (1394–1460) of the Kingdom of Portugal, began the European worldwide explorations. He is often referred to as the initiator of the Age of Discovery or Exploration, loosely timed as from early fifteenth century through the eighteenth century. Henry was a royal prince who started with the exploration of the coast of Africa, aiming especially at uncovering the sources of wealth in West Africa. His explorations also included the islands of the Atlantic Ocean and new maritime routes. He assisted the development of new and lighter sailing ships that expanded the range of his explorations. He and his successors developed an intimate knowledge and utilization of the wind conditions in the Mid-Atlantic. His observations aided eventually the discovery of the Americas.

Statue of Prince Henry the Navigator (José Simões de Almeida, 2002) on Belgrave Square, SW1, opposite West Halkin Street. It is a copy of the statue erected originally on São Miguel Island in 1932

Statue of Bartolomeu Dias at the South Africa House, Trafalgar Square, WC2N

Statue of Christopher Columbus (Tomás Bañuelos, 1992) on Belgrave Square, SW1, opposite the Spanish Embassy

Bartolomeu Dias (1450–1500) was a Portuguese explorer who charted the route from Europe to Asia. Documenting the first European to journey around the southernmost tip of Africa, in 1488, his recorded experiences were a great help to contemporary and future expeditions.

The Italian explorer Christopher Columbus (1451–1506)[2] sailed westward from Europe to reach India, sponsored by the Spanish throne in his endeavor. Instead of India, he reached the New World of the American continent. He made altogether four voyages across the Atlantic Ocean. His memorial was unveiled in Belgrave Square on the occasion of the 500th anniversary of his landing.

From left to right: Bas-relief of Sir Francis Drake by Henry Hugh Armstead or John Birnie Philip on the façade of the Foreign and Commonwealth Office; another bas-relief of Drake on the façade of the Visitors' Centre in Greenwich, King William Walk, SE10; and a plaque of Sir Francis Chichester, 9 St James's Place, SW1 (this image courtesy of Matt Brown)

---

[2]The pedestal of the statue gives Columbus's birth year mistakenly as 1446.

Sir Francis Drake (c.1540–1596) at the peak of his career was an admiral of the Royal Navy; at other times, he fought as a civilian under commission. Between 1577 and 1580, he became the second sea captain to circumnavigate the Earth. The first circumnavigation was by the Magellan-Elcano expedition between 1519 and 1522 from Seville, Spain. Drake and his fleet started from Plymouth and crossed the Atlantic, Pacific, and Indian Oceans. Though cruel and ruthless to the indigenous peoples he encountered, Drake earned much commendation from Queen Elizabeth I for his conquests.

Sir Francis Chichester was a twentieth-century avatar of Drake and Cook. He was a solo sailor in 1966–1967 when in nine months and one day he circumnavigated the globe. A memorial of three circumnavigators of the world was unveiled in 1979 in Westminster Abbey, in its south cloister: Sir Francis Drake, Captain James Cook, and Sir Francis Chichester.

Statue of Sir Walter Raleigh by William McMillan in Greenwich and his line engraving (Wellcome Collection)

Sir Walter Raleigh (also spelled Ralegh, 1552 or 1554–1618) was an explorer, but also accomplished in other realms of activity. Beside his statue in Greenwich, there is another in the North Carolina capital city named after him. Raleigh led expeditions to South America, though not to North America, yet his name is associated with North American British settlements. He popularized tobacco in England. He held high rank under Queen Elizabeth and was her favorite courtier, yet was also imprisoned in the Tower. Later he was freed and continued to serve in high offices. Under the next sovereign, James I, Raleigh was imprisoned again. During his long second imprisonment at the Tower, he wrote *The History of the World*, but it was left incomplete when he was beheaded. His burial place at St Margaret's Church, Westminster, is a World Heritage Site in a single unit together with the Palace of Westminster and Westminster Abbey. We are not aware of any scholarly publications by Sir Walter Raleigh, yet he is credited with having initiated one of the most famous mathematical problems. While preparing for one of his expeditions around 1587, he asked the mathematician and astronomer Thomas Harriot (c.1560–1621) about the most economical, i.e., the densest possible, means of packing cannon balls in his ship. It was decades later that Johannes Kepler gave the solution, but without proof. The proof was finally provided only toward the end of the twentieth century.

Left: Statue of John Smith, a copy by Charles Rennick, 1960 (of the original by William Couper, 1907, in Jamestown, Virginia), Bow Churchyard, EC4. Right: Portrait of William Dampier by Thomas Murray, 1697–1698 (Wikimedia)

John Smith (1580–1631) was a controversial explorer and soldier. He attended school between 1592 and 1595, but after his father died, he began his seafarer and military career. He was a mercenary, a pirate, a defender against pirates, and a cavalry officer in continental wars; was ennobled by a central European sovereign, Sigismund Báthory, Prince of Transylvania; and traded in slaves. In 1604, he returned to England and started his career as a colonist in North America. He played a pivotal role not only in establishing the first permanent settlement, the Jamestown colony, but also in ensuring its survival, training the colonists in self-sufficiency. As an explorer, he led an expedition to what is today Virginia and the Chesapeake Bay and mapped the Chesapeake Bay area. He produced other maps, published books, and helped colonizing New England, a name he coined. He died in London.

William Dampier (1651–1715) circumnavigated the Earth three times and was an influential figure bridging the era between Sir Walter Raleigh and Captain James Cook. Several geographical locations have been named after him in Australia, New Zealand, and elsewhere. He introduced many words into the English language, among them such commonly used ones as *banana*, *barbecue*, and *chopsticks*. His accounts of his explorations helped later expeditions and inspired writers, among them Jonathan Swift in producing *Gulliver's Travels*. There is a monument honoring him in the Bedford Memorial Park in Broome, Western Australia. It was built in the form of a sea chest and was unveiled in 1938.

Australia

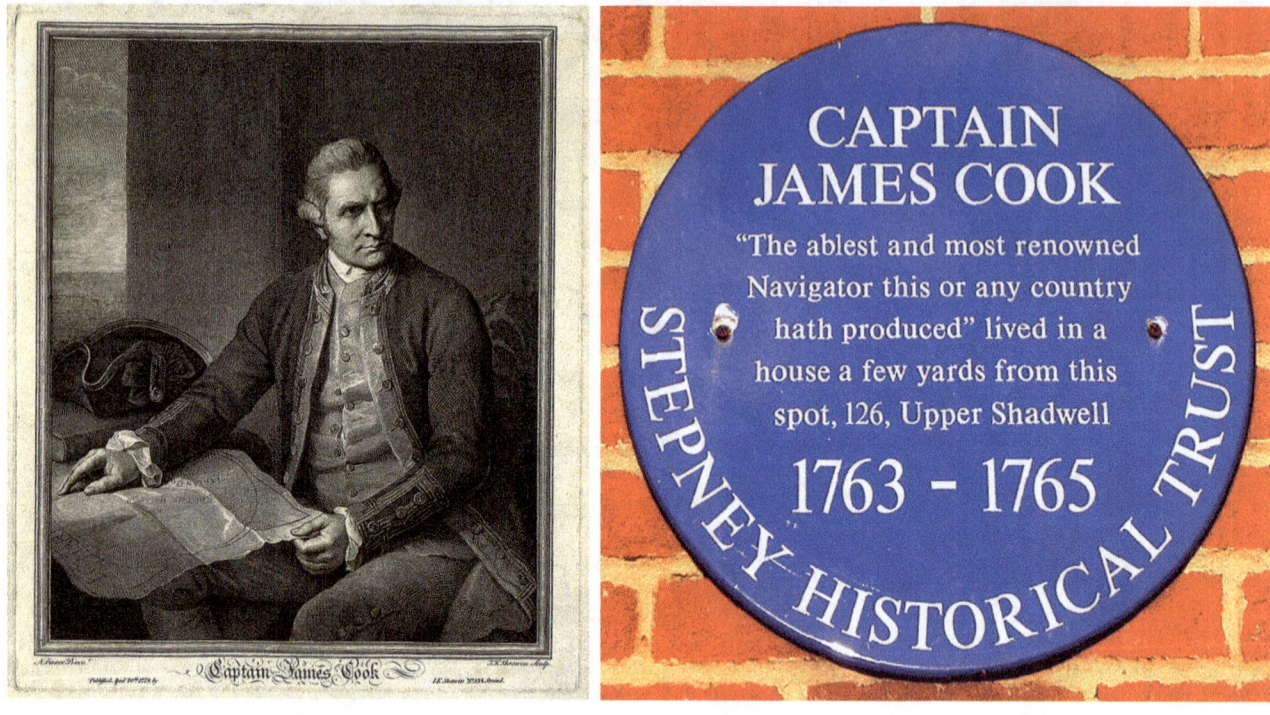

Captain James Cook: Line engraving by J. F. Bolt, 1788, after Sir N. Dance-Holland, 1776 (Wellcome Collection), and a plaque, 326 The Highway, E1, close to the place where he lived (courtesy of Matt Brown)

Bas-reliefs of Captain James Cook: left, by Henry Hugh Armstead or John Birnie Philip, on the façade of the Foreign and Commonwealth Office, and right, on the façade of the Visitors' Centre in Greenwich, King William Walk, SE10

James Cook (1728–1779) was a navigator who rose from very modest origins to world fame. After he joined the Navy in 1755, he rose quickly to the rank of captain. He participated in surveying the St Lawrence River in Canada and the shores of Newfoundland. He commanded the vessel *Endeavour* for the expedition of the Royal Society to the Pacific and in particular Australia and New Zealand. He was also a cartographer and geographer and contributed greatly to the knowledge about the Pacific and the Southeastern Ocean. He was killed in Hawaii, where today his statue commemorates his explorations—just one among the many memorials honoring him.

Statue of Captain Matthew Flinders by Mark Richards at Euston Station, NW1, and his plaque, 52 Fitzroy Street, W1

Captain Matthew Flinders (1774–1814) was another explorer intimately connected with Australia. He was the second who circumnavigated this continent. The first was the Dutch Abel Tasman in 1642–1643, who called the massive island "New Holland." The name *Terra Australis*, Latin for "southern land," had been used for a supposed southern continent. Flinders simplified it to *Australia*. As a navigator and geographer, Flinders conducted scientific explorations. When he was returning from his third voyage to the south to England in 1803, he had to stop for repairs at what is today the Republic of Mauritius, then a French colony, an island in the Indian Ocean. At the time, Britain and France were at war; as a consequence, the French governor arrested Flinders and detained him for 6 years. During his captivity, Flinders prepared the manuscript with maps of his monograph, *A Voyage to Terra Australis*. While he labored on this publication, he was severely ill; on the day of publication of the monograph, he died. Since his death there has been a widespread cult of Flinders in Australia, with large statues and other memorials second only to those of Queen Victoria.

Africa

LIVINGSTONE

David Livingstone's statue by Thomas Bayliss Huxley-Jones, 1953, on the front façade of the Royal Geographical Society headquarters, facing Exhibition Road and his bas-relief on the façade of the Foreign and Commonwealth Office, Whitehall, SW1, by Henry Hugh Armstead or John Birnie Philip

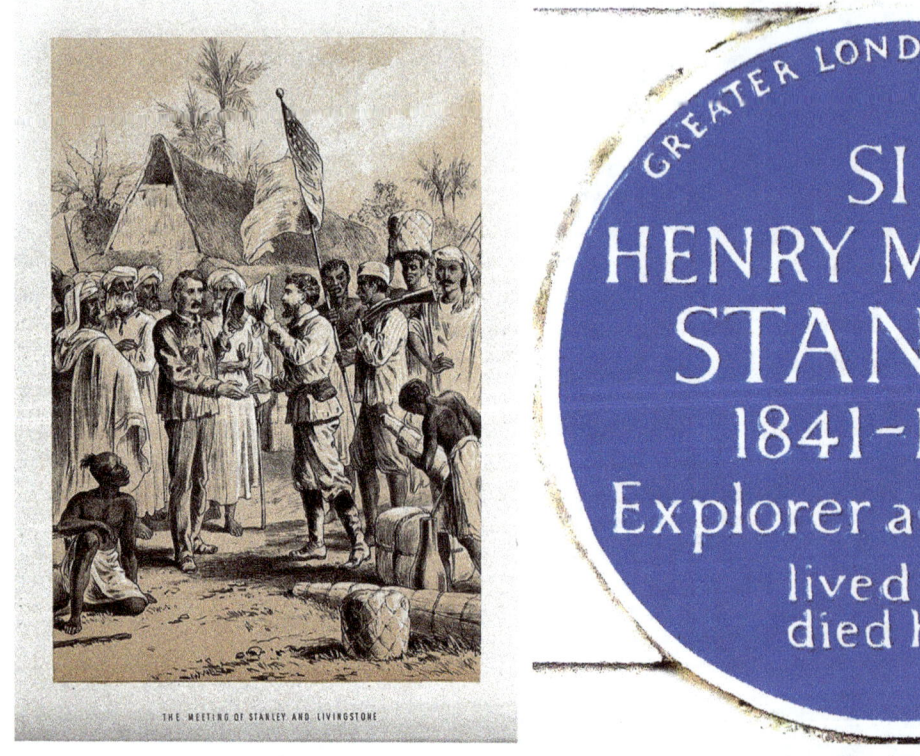

Henry Morton Stanley meeting David Livingstone in Central Africa. Lithograph (Wellcome Collection). Sir Henry Morton Stanley's plaque, 2 Richmond Terrace, SW1A (Spudgun67)

David Livingstone (1813–1873) was born in Blantyre, Scotland, into a family of laborers. Despite financial struggles, he obtained his first medical qualification from the Faculty of Physicians and Surgeons of Glasgow. He continued his medical studies in London in preparation for missionary work. While ostensibly a medical missionary, as an explorer he was looking for the sources of the Nile River. His ultimate moral goal was to end slave trade in East Africa. Ironically, he was more successful in geographical endeavors than in his missionary activities, claiming the rights to name many lakes and rivers, among them Victoria Falls. He traveled a great deal, was severely ill yet persevered, and carried on under the most trying conditions. When Livingstone disappeared to the outside world in 1869, the *New York Herald* provisioned a journalist, Henry M. Stanley (1841–1904), to find him. Stanley eventually also became a noted Africa explorer. He located Livingstone in 1871 on the shores of Lake Tanganyika. Upon their meeting, supposedly, they had the following exchange. Stanley: "Dr. Livingstone, I presume?" Livingstone: "Yes. I feel thankful that I am here to welcome you." This exchange sounds too cool under the circumstances; besides, Stanley could have had no doubt that he had found Livingstone. Despite its implausibility, this legend has received wide popularity. Livingstone declined to leave Africa and died soon after being contacted by Stanley. He is interned at Westminster Abbey (except his heart, which was buried where he died, in Africa). Memorials honor him in Britain, in Africa, and elsewhere.

Sir Richard Francis Burton's portrait photograph (Wellcome Collection) and his Mausoleum by Isabel Burton, photograph by Svarochek (Creative Commons)

Sir Richard Francis Burton (1821–1890) was an explorer and a comrade of John Hanning Speke, but they became alienated due to their fierce disagreement over the location of the source of the River Nile. Burton explored Asia, Africa, and the Americas, spoke 29 languages, and had an encyclopedic knowledge on most subjects. As a linguist, his most lasting contribution was a well-regarded translation of *The Arabian Nights*. Although a member of the Anglican Church, he was an atheist. His widow was a devout Catholic who had his scientific writings destroyed, but had a mausoleum designed for him. It is the famous Tent Tomb Memorial at the graveyard of St Mary Magdalene Church, 61 North Worple Way, SW14.

Left: John Hanning Speke Obelisk, 1866, in Kensington Gardens. Right: Memorial of Frederick C. Selous by William Robert Colton in the wall, on the stairway to the first floor on the left, in the main hall of the Natural History Museum

John Hanning Speke (1827–1864) was an explorer of Africa. His best-known search was for the source of the Nile. Only in the 1870s was it shown that he had surmised correctly that the Nile flowed from Lake Victoria. Speke conducted three expeditions to Africa. His valuable observations were published in 1863.

Frederick Courteney Selous (1851–1917) started collecting items relevant to natural history from early childhood; this habit combined with his interest in explorations. David Livingstone was one of his role models. Selous was educated in England, Germany, and Austria. First traveling to Africa when he was 19 years old, he became most knowledgeable about this continent and traveled throughout it extensively, doing ethnological investigations. He joined the British South Africa Company in 1890, helped extend the British rule of that part of the continent, and conducted broad-scale explorations. The Royal Geographical Society awarded him its Founder's Medal. He earned fame as both a conservationist and big game hunter; beside Africa he made extensive hunting trips to Europe, Asia, and North America. The conservationist US president, Theodore Roosevelt, was among his friends. Selous amassed a huge collection of diverse items important for natural history and it has enriched museums and collections, in particular the Natural History Museum in London. Later in his life, Selous entered British military service and he fell in fighting German troops in Tanzania.

Plaque of Mary Kingsley, 22 Southwood Lane, N6 (Spudgun67), and her portrait by A. G. Dew-Smith (Wikimedia)

Mary Kingsley (1862–1900) had little formal education as even her educated parents did not consider it necessary for girls to have equal advantages. However, she made good use of her father's rich library. She was especially interested in reading about science and about explorers. When her parents died, she used her inheritance to travel extensively in West Africa. She had fruitful interactions with the British Museum and with a publisher that was willing to bring out her planned book. She persevered in spite of the difficulties a single female traveler faced. Upon return to England, she produced two books about her experiences, and they were met with interest as were her numerous lectures for broad audiences. When the second Boer War broke out, she traveled to Cape Town and treated wounded prisoners of war in a hospital. There, she contracted typhoid fever and died. As she had wished, she was buried at sea.

Polar Exploration

Statue of Sir John Franklin by Matthew Noble, 1866, at the western side of Waterloo Place, SW1, and the bronze relief at the front of the plinth depicting Franklin's funeral

From left to right: Bas-relief of Sir John Franklin by Henry Hugh Armstead or John Birnie Philip on the façade of the Foreign and Commonwealth Office and his bust at the Maritime Museum in Greenwich, Park Row, SE10

Sir John Franklin (1786–1847) early on sought a career on the sea. Joining the Royal Navy as a teenager, he saw much combat. Eventually he led expeditions such as the Coppermine Expedition of 1819–1822 from the Hudson Bay to chart the north coast of Canada. It ended under cruel conditions, but the experience did not deter Franklin from continuing to conduct expeditions to the North of Canada and the Arctic regions. He made significant contributions to charting the Arctic coastline. In 1845, an expedition consisting of two ships, the *Erebus* and the *Terror*, was sent under Franklin's command to determine the remaining portion of the Northwest Passage. The two ships became trapped in ice in fall 1846 and Franklin died there. The lost ships were not found until the 2010s, the wreck of *Erebus* in 2014 and that of *Terror* in 2016.

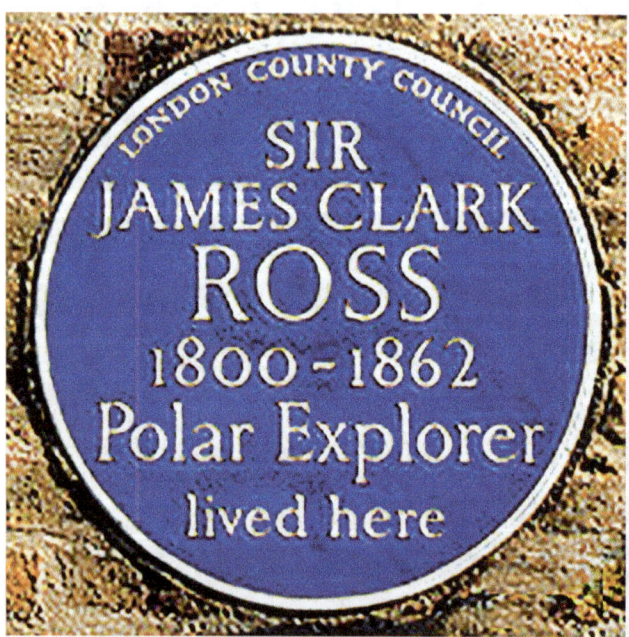 

Left: Plaque of Sir James Clark Ross, 2 Eliot Place, SE3 (Spudgun67). Right: Plaque of John Rae, 4 Lower Addison Gardens, W14 (Spudgun67)

Sir James Clark Ross (1800–1862) participated in Arctic expeditions with his uncle, Sir John Ross (1777–1856) and with Sir William Edward Parry (1790–1855). Later, his attention turned to Antarctica; he led his own expedition there between 1839 and 1843. The naval hydrographer Francis Beaufort provided a broader scientific background for the project. The expedition discovered several new regions, some of which today carry Ross's name.

John Rae (1813–1893) studied medicine at the University of Edinburgh and became a surgeon licensed by the Royal College of Surgeons of Edinburgh. Later, as part of his assignment in exploring northern Canada, he learned surveying. Although Rae's provisions were meager in comparison with those of naval officers who led similar expeditions, Rae persevered. He made several journeys to northern Canada and contributed much to charting this heretofore unknown region and to establishing the Northwest Passage. While looking for the lost expedition of Sir John Franklin, Rae found some evidence of its fate. The recognition of Rae's achievements came only recently. The blue plaque shown above was established in 2011; his statue in his birthplace, in Orkney, was unveiled in 2013, and a memorial plate in his honor was installed at Westminster Abbey in 2014.

Joseph R. Bellot Obelisk in Greenwich

Joseph R. Bellot (1826–1853) was a French Arctic explorer who has a memorial obelisk in Greenwich in front of Greenwich Hospital. He participated in British expeditions to Madagascar, to South America, and to the Arctic. They were searching for John Franklin in one of the Arctic expeditions. The young Bellot perished in another Arctic expedition.

Statue of Robert Falcon Scott by Kathleen Scott (his widow), 1915, Waterloo Place, SW1, and his plaque, 56 Oakley Street, SW3 (Spudgun67)

Robert Falcon Scott (1868–1912) was a British naval officer whose chance encounter in 1899 with Clements Markham, the President of the Royal Geographical Society, changed his life. Markham told Scott about an Antarctic expedition then forming. Scott led the Discovery Expedition, 1901–1904. In the first year, in a final though unsuccessful attempt to reach the South Pole, Ernest Shackleton and Edward Adrian Wilson accompanied Scott. The next year, the expedition discovered the Polar Plateau and collected a wealth of biological and geological data. Shackleton and some others did not stay to the completion of the expedition. Soon, Shackleton and Scott became rivals in polar exploration. Shackleton led another Antarctic expedition, but did not quite reach the South Pole.

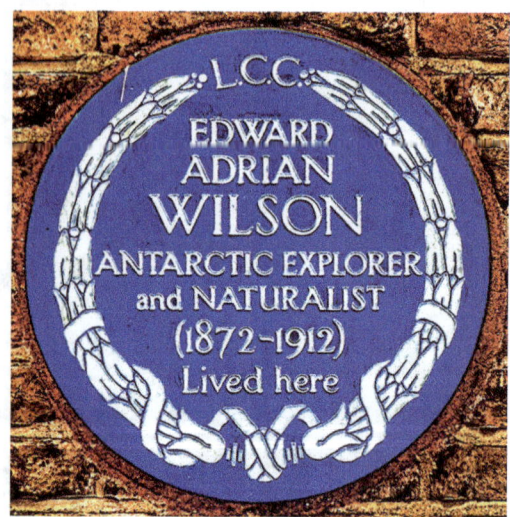

Plaque of Edward Adrian Wilson, 42 Vicarage Crescent, SW11 (Spudgun67)

From left to right: Lawrence Oates, Henry Robertson Bowers, Robert Falcon Scott, Edward Adrian Wilson, and Edgar Evans (Wikimedia)

Scott embarked on his fateful expedition, called Terra Nova, in 1910. Five of his men went for the final stretch to the Pole: in addition to Scott, Henry Robertson Bowers (1883–1912), a naval officer and photographer; Edgar Evans (1876–1912), a polar explorer; Lawrence Oates (1880–1912), an army officer and Antarctic explorer; and Edward Adrian Wilson (1872–1912), a polar explorer, physician, and ornithologist. They reached the South Pole on January 17, 1912, only to learn that five weeks before, the Norwegian Roald Amundsen and his four associates had just beat them to the prize on December 14, 1911. Tragically, on their way back, Scott and his four associates perished.

Sir Ernest Henry Shackleton statue by Charles Sargeant Jagger (the head based on J. A. Stevenson's bust), on the front façade of the Royal Geographical Society headquarters, facing Exhibition Road, and his plaque, 12 Westwood Hill, SE26 (Spudgun67)

Sir Ernest H. Shackleton (1874–1922) participated in Robert Falcon Scott's Discovery Expedition, 1901–1904, to Antarctica, but had to return home in 1903, before the completion of the mission on account of ill health and his conflicts with Scott. In 1907–1909, he led his own Nimrod Expedition and approached the South Pole closer than anybody had before. In the next, Imperial-Antarctic Expedition, 1914–1917, he was aiming at crossing Antarctica, from sea to sea. However, their ship *Endurance* was trapped in ice, so Shackleton and his crew mounted a daring escape. Perhaps, the name of their ship expressed best his most characteristic trait, which made him a role model for many. He died of heart attack during yet another Antarctic expedition.

Scientists as Explorers

Exploration and science went often hand in hand to varying degrees. Here we single out three explorers whose science-related activities were especially noteworthy.

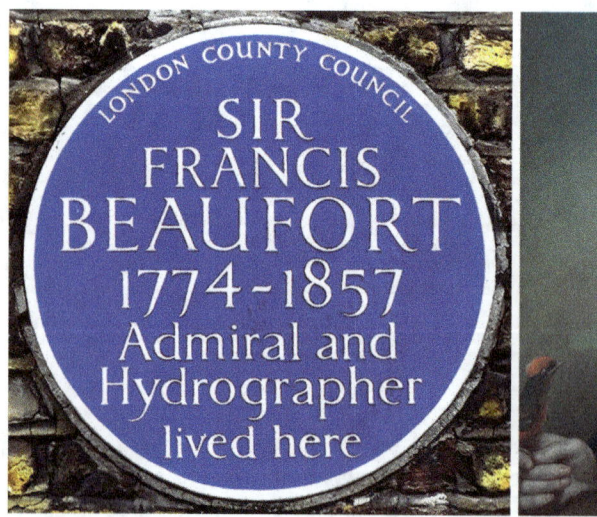

Plaque of Sir Francis Beaufort, 52 Manchester Street, W1, and portrait of Charles Waterton by Charles Willson Peale, 1824 (Creative Commons)

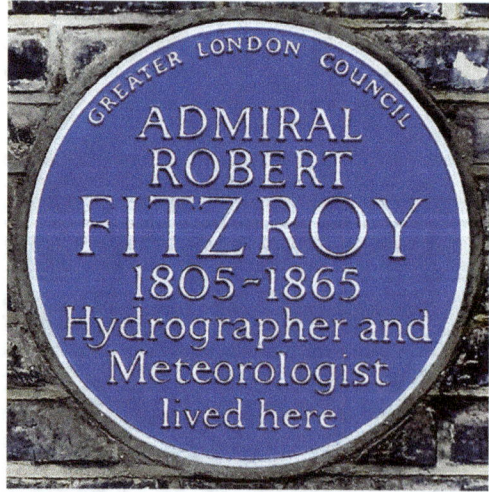

Plaque of Robert Fitzroy, 38, Onslow Square, SW7; and his portrait (Wellcome Collection)

Sir Francis Beaufort (1774–1857) is best known for the Beaufort scale as a measure of wind force. He left school when he was 14 to become a seaman. However, he went on with self-education for the rest of his life and associated with great scientists, such as John Herschel, Charles Babbage, and Sir George Airy. Eventually, he received a captain's commission in the Navy. He used his off-duty time for scientific activities, made astronomical observations, and charted maps. As his interests and activities became known, the Admiralty asked him to conduct hydrographic surveys in South America and Anatolia. These surveys involved all kinds of measurements and description that affect navigation, marine construction, and even oil drilling. At the age of 55, when most in the Navy would retire, he was appointed the British Admiralty Hydrographer. He held administrative duties for the astronomical observatories at Greenwich and Cape of Good Hope in Africa, directed maritime explorations, and was placed in command of the search for Sir John Franklin who was lost while looking for the Northwest Passage. It was at Beaufort's initiative that Charles Darwin was invited to participate in the second voyage of the *Beagle* that contributed so much to that naturalist's milestone discoveries. Beaufort had trained the *Beagle's* Captain Robert Fitzroy, and Fitzroy asked him for somebody well-educated and scientific to serve as ship's naturalist. Beaufort's support was crucial for Darwin's further science-related projects and expeditions.

Charles Waterton (1782–1865) was an explorer, especially of South America. His accounts of his explorations inspired the next generation to follow in his footsteps, among them Charles Darwin and Alfred Russel Wallace. One of Waterton's disciples even taught the young Darwin taxidermy. Waterton, a pioneer conservationist, walled off a large chunk of his estate to transform it into a nature preserve. For his concern with air pollution, he is recognized as a pioneer environmentalist.

Robert Fitzroy (1805–1865), admiral and scientist, is best known for having been the captain of the second voyage of the *Beagle* with Charles Darwin and from his explorations of Tierra del Fuego and the Southern Cone. Fitzroy, who had an aristocratic background, enrolled in the Royal Naval College in Portsmouth when he was 12 years old, and at 14 he was already on an expedition to South America. He was appointed captain of the *Beagle* in 1828 on a temporary basis. He obtained its permanent command at the intervention of his former mentor, Francis Beaufort. Upon the return of the *Beagle* from its second voyage, Fitzroy published a four-volume treatise of the expedition with much useful information for future travelers to the region. The third volume of this set was Darwin's journal.

Fitzroy's later career included his governorship of New Zealand, 1843–1845. He retired from active service in 1850, but the rest of his life remained significant for science. After he was elected Fellow of the Royal Society with Darwin's backing, Fitzroy was appointed chief of a new department of Britain, which was charged with collecting the data on weather. It was the predecessor of the present-day Meteorological Office (Met Office since 2000). Fitzroy took his new assignment seriously. A new type of barometer he distributed to every port made accurate weather information available to all crews before embarking on lengthy voyages or local fishing runs. He encouraged work on perfecting the barometer and introduced a system of warning for storms and gales and instituted measures that would keep fleets in port if sailing was unsafe. Queen Victoria appreciated his achievements. For all the progress Fitzroy helped to happen, he was unhappy to see Darwin's *On the Origin of Species* and its success. As he perceived it, Darwin's premise went against his religious beliefs. When various troubles hindered the proper operations of his Meteorological Office, he committed suicide.

Sir Clements Robert Markham bust at the back entrance of the Royal Geographical Society, 1 Kensington Gore, SW7, and his photograph (Wellcome Collection)

Sir Clements Robert Markham (1830–1916) had a rich and at times controversial career as an explorer and as a society administrator. Early on he was a crew member of the ship *Assistance* in one of the search operations for John Franklin's lost expedition. Markham studied the history, geography, and archeology of Peru in two expeditions. His bust was a gift from the Peruvian government as an expression of gratitude for Markham's services as a historian of Peru. On his second mission, he collected seeds of the cinchona tree to be sent to India. The seeds were planted there to grow the valuable tree, whose bark was a source of quinine, the first known remedy against malaria and other tropical diseases. Markham was active in launching expeditions, especially to the Antarctic region, and in administering societies. He was secretary of the Royal Geographical Society for 25 years and president for 12 years. In his later years, he wrote history, biographies, and travel accounts.

Greenwich

Many maritime voyages of discovery started from Greenwich. Its Maritime Museum and the Royal Observatory signify this history and the pioneering activities of British astronomers. Their numerous discoveries are commemorated, among them the continuous improvement of instrumentation for investigating the heavens and bringing centuries-old human dreams into human proximity.

Visitors' Centre. Its façade depicts as busts 13 great naval leaders, among them two explorers (Cook and Drake) presented above. The statue of Sir Walter Raleigh stands in front of the building (see above)

Flamsteed House of the Royal Observatory and the portrait of its designer, Sir Christopher Wren, by E. Scriven, 1809, after M. Rysbrack (Wellcome Foundation)

The Flamsteed House is the oldest structure of the Royal Observatory. It was completed in 1676, following the design of Sir Christopher Wren (1632–1723) who was not only an architect but also a renowned astronomer (Chap. 3). The construction was supervised by Robert Hooke (Chap. 3). Wren is best known as an architect, but he was previously Professor of Astronomy at Oxford University. Flamsteed House received its name after its first inhabitant, the first Astronomer Royal, John Flamsteed. Edmond Halley and George Airy were among its subsequent inhabitants.

Further structures of the Royal Observatory

Bust of John Flamsteed by J. Raymond Smith on the façade of the Royal Observatory and Flamsteed's portrait (Wellcome Collection)

John Flamsteed (1646–1719) studied at Derby School and continued his studies at home as an extended Cambridge sojourn proved impossible for him due to illness. Still, he visited Cambridge and attended Isaac Newton's Lucasian Lectures. Flamsteed became interested in astronomy as a teenager and wrote his first scientific paper when he was 19. He became involved with establishing the Royal Observatory and he was the first Astronomer Royal although initially his position was called "The King's Astronomical Observator." The actual work on building the Royal Greenwich Observatory commenced in 1675. The following year he moved into the Observatory, where he lived for 8 years. In 1684 he took up the position of rector in a small village, but he carried on as Astronomer Royal until the end of his life. His principal work was a star catalogue, *Catalogus Britannicus*, with close to 3,000 entries and an atlas of the stars, *Atlas Coelestis*. Both appeared after his death. His wife, Margaret, initiated their publication and participated in the necessary editorial work.

Left: Edmond Halley, mezzotint by J. Faber, 1722, after T. Murray, 1712 (Wellcome Collection), and the mark of Halley's meridian line of 1725 at the Royal Observatory. The text on the rod reads vertically, "HALLEY'S MERIDIAN LINE." Right: Portrait of Sir George Biddell Airy (Wellcome Collection)

Edmond Halley (1656–1742) studied at Queen's College in Oxford and became interested in astronomy during his student years. He was in contact with some of the great scientists of his time, such as John Flamsteed, Robert Hooke, and Christopher Wren. He traveled throughout Europe making astronomical observations: he calculated the orbits of comets, charted terrestrial magnetism, and observed the eclipse of the Sun in 1715 in London. After the death of Flamsteed, he was appointed the second Astronomer Royal (1720–1742) and predicted that the comet that now bears his name would return in 1758. Among other functions and involvements, he was secretary of the Royal Society and supported the publication of *Principia*, Isaac Newton's seminal treatise.

Halley's meridian line in Greenwich was established as the prime meridian in 1851 by Sir George Biddell Airy (1801–1892), who was Astronomer Royal between 1835 and 1881. The movement to accept the Greenwich meridian as the official meridian gained momentum in 1884. This followed the initiative of US President Chester A. Arthur, who hosted the International Meridian Conference in Washington, DC (the abstaining France joined the rest a few decades later).

An Edmond Halley memorial (not shown here) by Richard Kindersley in the south cloister of Westminster Abbey was unveiled in 1986, the year when the Halley comet once again returned. The memorial makes note of the fact that on this visitation, the path of the comet was intercepted by the spacecraft *Giotto*. The monument features a miniature image of *Giotto* in the center of the Halley's portrait. The monument itself is in the stylized shape of a comet, while the gilded tail lists some of Halley's achievements.

The remains of Herschel's telescope in the garden of the Royal Observatory and how it looked when it was operational; 40-foot telescope constructed by William Herschel, for use outdoors, colored etching; and Sir William Herschel and Caroline Herschel. Color lithograph by Alfred Richard Diethe, ca. 1896. They are polishing a lens or a mirror of their telescope (all, Wellcome Collection)

Sir Frederick William Herschel (1738–1822) was a German-born British astronomer who designed instrumentation and made discoveries. His 40-foot (12 meter) telescope was the largest in the world at the time. He discovered Uranus in 1781 using this telescope. Herschel was also a pioneer in utilizing spectrophotometry and discovered infrared radiation. These were of seminal importance for astronomy, but also in other fields especially for the investigation of the properties of matter. He was also a well-trained and accomplished composer. His younger sister, Caroline Lucretia Herschel (1750–1848), was also an internationally renowned astronomer. She worked with her brother throughout her life, but made independent discoveries as well. Her career opened the way for future women scientists. She was the first woman to hold a government position in England. She was earning a salary for doing science at the time when even men seldom were paid for similar work. She was the first woman to receive a high recognition—this was in 1828—from the Royal Astronomical Society. In 1835, she was named an Honorary Member of the Society; in this first she shared the distinction with Mary Somerville. Caroline Herschel's best-known achievements in astronomy were her discoveries of several comets, which she published in the period 1782–1787. The Royal Family was interested in her discoveries and invited her brother to give an account of them in Windsor Castle. Apparently, the Royals did not want to go as far as getting her out to Windsor. After her brother's death, she continued her research.

 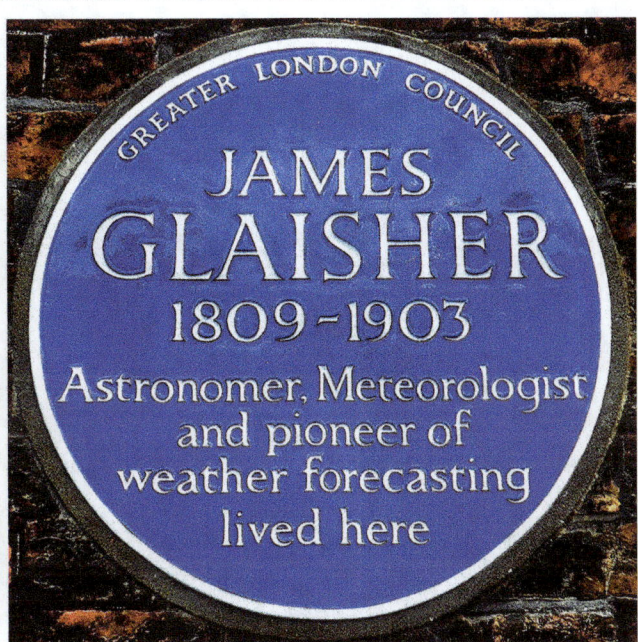

Plaques of Luke Howard, 7 Bruce Grove, N17 (left), and James Glaisher, 20 Dartmouth Hill, SE10 (right, both, Spudgun67)

Luke Howard (1772–1864) was a pharmacist by training who later operated a successful factory producing industrial chemicals and pharmaceuticals. He was also an amateur meteorologist but so good at his hobby that he earned the label "the father of meteorology." He made detailed observations of clouds and classified them, coining names for their principal categories. Before him, the great French naturalist Jean-Baptiste Lamarck (1744–1829) had described cloud formations. However, Howard's Latin names and his additional description of transitional forms made his system more popular and more broadly applied than Lamarck's.

James Glaisher (1809–1903) was a meteorologist and astronomer. Following a period at the Cambridge Observatory, in 1835 he joined the Royal Observatory in Greenwich. For 34 years, he served as the Superintendent of the Department of Meteorology and Magnetism. He is called an aeronaut for his pioneering studies of the atmosphere. He made numerous ascents with balloons to measure the temperature and humidity at high altitudes. He broke the world record of altitude as he may have reached at least 9500 meters, perhaps higher, but he passed out on that occasion and could not record measurements.

Plaques of Sir Frank Dyson, 6 Vanbrugh Hill, SE3, in Greenwich (left), and Will Hay, 45 The Chase, SW16 (right, both, Spudgun67)

Sir Frank Dyson (1868–1939) was Astronomer Royal between 1910 and 1933. He started his career in Greenwich in 1894. His main interest was in solar eclipses, and he participated in three expeditions between 1900 and 1905 in order to observe the corona and chromosphere of the Sun. He personally organized yet another expedition to observe the 1919 solar eclipse. He reported his observations to the joint meeting of the Royal Society and the Royal Astronomical Society on November 6, 1919. It was solid evidence of the effect of gravitation on light, hereby confirming Einstein's theory of general relativity (see more in connection with Sir Arthur Stanley Eddington in Chap. 3).

Will Hay (1888–1949) was an actor popular and successful on both stage and film. He was also an amateur astronomer with no less dedication than any of his professional peers. He built his private observatory, constructed instruments, and built them with his own hands. He observed a temporary white spot on the planet Saturn, measured the positions of comets, gave lectures, and published a book (the last two under the name of W. T. Hay).

3
Scientists

Statue of Charles Darwin by Sir Joseph Boehm, 1885, in the main hall of the Natural History Museum

"Geometry" and "Philosophy" by John Birnie Philip, 1868, and "Astronomy" and "Chemistry" by Henry Hugh Armstead, 1868, all four at the Albert Memorial, Kensington Gardens, W2

Today it may appear strange that physics was not represented by an allegorical statue by the Victorians, whereas geometry, philosophy, astronomy, and chemistry were. At the time, however, much of what we would today call physics was labeled natural philosophy. Even today, physicists and other natural scientists earn the title of Doctors of Philosophy, not just the bona fide philosophers and fellow humanists.

The former building of the City of London School for boys on Victoria Embankment next to the Millennium Bridge for pedestrians, opposite Tate Modern

Allegorical representation of Geometry, Mathematics, Chemistry, and Mechanics by G. W. Seale in spandrels on the façade of the former building of the City of London School

The City of London School was founded in 1442. The late Victorian building shown here housed the School between 1883 and 1986. There are four statues by J. Daymound & Sons on the front of the building and ten allegorical figures representing fields of learning that the School found important enough for such emphasis. Facing the front of the building from left to right this is the order of the decorations: Geometry, Mathematics, Francis BACON (Chap. 1), Drawing, Music, SHAKESPEARE, Classics, Poetry, MILTON, Sacred History, Ancient History, NEWTON (see below), Chemistry, Mechanics. There is another statue on the side of the building, Sir Thomas More's.

John Harvard (1607–1638) was one of the graduates of the City of London School. He served as clergyman in Colonial America. His bequest for the recently founded school at the Massachusetts Bay Colony prompted the authorities of this institution to name it after him; hence, the name of Harvard University.

Royal Society

Two views of the Headquarters of the Royal Society, 6–9 Carlton House Terrace, SW1Y, from the Carlton House Terrace (left) and from The Mall (right). The monument on the left is the Duke of York Column

An "invisible college," a group of physicians and natural philosophers, congregated around Robert Boyle (1627–1691) in the middle of the seventeenth century in London. These learned men felt an increasing need to come together from time to time to exchange their views and discuss their observations (more about him below). As similar groups formed around the country, the time appeared ripe for creating a more formal institution for such gatherings in London. On November 28, 1660, Christopher Wren, Professor of Astronomy at Gresham College, gave a lecture at the College and convened a meeting after the lecture. This meeting may be considered as the initial founding of the Royal Society.

Statue of Sir Thomas Gresham by Henry Bursill, 1868–1869, in a niche of Gresham House, south-east pavilion, Holborn Viaduct, EC1, and the east end of the Royal Exchange with its clock tower beneath which is another Gresham statue by William Behnes, 1844–1845, partially visible from this view

Sir Thomas Gresham (c.1519–1579) was a merchant and financier whose career overlapped with the reigns of three monarchs: King Edward VI, Queen Mary I, and Queen Elizabeth I. As a financial expert he served all three monarchs, and his best-known deed was the founding of the Royal Exchange. A grasshopper motif decorating the top of the clock tower of the Royal Exchange and elsewhere on the façade of the building references Gresham's coat of arms. According to the family legend, when one of Gresham's ancestors was lost as a baby, the sound of a grasshopper alerted a servant woman, who then rescued him.

Current venue of Gresham College, Barnard's Inn on Holborn, EC1N, between Furnival Street and Fetter Lane

Sir Thomas Gresham left money in his will to establish an institution that was founded in 1597 as Gresham College, the first institution of higher education in London. Its purpose was to bring new knowledge to the public through lectures, which townspeople could attend free of charge. This mission is still in effect, and the College continues a rich lecture program. Today, the public lyceum is its only activity. Its current location is in Holborn, but some lectures are held at other locations, among them the Museum of London.

When the Society was formed in 1597 it was not yet called Royal. That came later when King Charles II granted it a royal charter. It was decided at the first meeting by the 12 founders that their mission would be the extension of knowledge by means of *experimentation*. This was expressed in the motto of the Society, *Nullius in verba*, in English, "Take nobody's word for it." With this as their watchword, they would come together, in person, to present and discuss their observations. They would accept no authority; only facts. Soon they had a royal charter signed by King Charles II, and the world's first national science academy came into existence. The list of original members is remarkably diverse not only in professional backgrounds but also in religion and politics; they included royalists and parliamentarians alike. As for their professions, they were astronomers, mathematicians, inventors, physicians, and natural philosophers—both professionals and amateurs.

Today the Royal Society has over 1,600 Fellows, among them over 60 Nobel laureates. Its alumni number over 8,000 Fellows, among them 280 Nobel laureates. In the United Kingdom and the British Commonwealth, the membership, i.e., being a Fellow of the Royal Society (FRS), is the most coveted recognition among scientists. It still has tremendous prestige internationally. The Society elected the first women as Fellows in 1945. The Headquarters of the Royal Society has been at 6–9 Carlton House Terrace since 1967. Before that date, its location was at Burlington House. There are two national academies adjacent to the Royal Society on Carlton House Terrace, the Royal Academy of Engineering and the British Academy, which is the national academy for the humanities and social sciences.

In the lobby of the Carlton House Terrace headquarters, there are a few busts and paintings.

From left to right: Portraits of John Wilkins by Mary Beale (Wellcome Collection), Seth Ward by John Greenhill (Wikimedia), and William Brouncker, from a painting by Sir Peter Lely at the Royal Society (Wellcome Collection)

John Wilkins (1614–1672) was a clergyman, a scientist, and one of the founders of the Royal Society. He was educated at Oxford and eventually became the head of an Oxford college and of a Cambridge college—this was Wilkins's unique distinction. He mentored Christopher Wren, among others, and steered the young Wren's interest toward science. As evidenced by his university ecumenism, Wilkins had a great ability to bring together people of sharply different views. At his time there was strong partisanship of royalists on the one hand and parliamentarians on the other. Yet when the Royal Society was established, both contingents participated in a bipartisan manner. Furthermore, Wilkins initiated the so-called natural theology, which advocates the compatibility of believing in the existence of God, but basing this belief on observation and of experience of nature. Wilkins's advocacy of "peaceful coexistence" may have contributed more to British scientific life at the time than the output of his science.

Seth Ward (1617–1689) was a mathematician and astronomer educated at Cambridge University. In 1649 he was appointed to be Professor at Oxford University. He was one of the original members of the Royal Society. Eventually, he moved to important church positions and became Bishop of Salisbury. There is a painting of Seth Ward by John Greenhill in the lobby of the Society.

William Brouncker (1620–1684) received medical training at Oxford University and was engaged in mathematical studies in which he achieved considerable success. One of the original members of the Royal Society, he served as its president from the beginning to 1677. Later he was the head of St Catherine Hospital.

Bust of King Charles II by Joseph Nollekens, 1780, in the lobby of the Royal Society

King Charles II (1630–1685) reigned from 1660 to 1685 and chartered the Royal Society early in his reign. He was the son of King Charles I who was executed in 1649. Charles II lived in exile until, following Oliver Cromwell's death in 1658, he was called back to the throne. Not all his biographies mention his role in establishing the Royal Society, but for our story, this was significant.

We single out three individuals for their role in the early years of the Royal Society.

 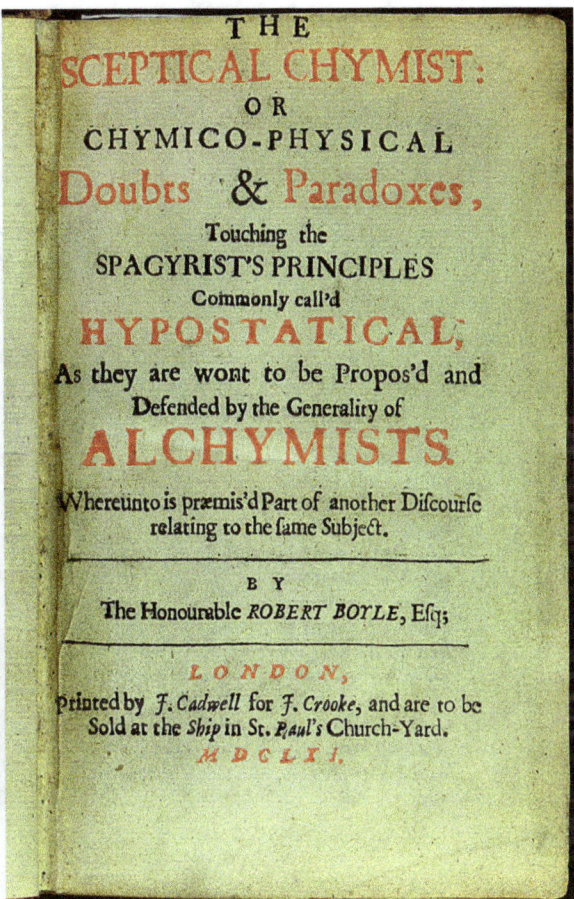

Portrait of Robert Boyle and his book, *The Sceptical Chymist* (both, Wellcome Collection)

Robert Boyle (1627–1691) was born in Ireland and was first educated at home. At the age of 8, he was sent to Eton, and at the age of 11, he and his brother were sent to Geneva with a French tutor. After 2 years, he continued his schooling in Italy, and he returned home when he was 15. Then he lived in Dorsetshire, England, until 1654, when he moved to Oxford as he was devoting his life to science. In Oxford he was associated with a society of men who called themselves the Philosophical College. It moved to London and was one of the forerunners of the Royal Society of London. While still in Oxford, Boyle hired Robert Hooke as his assistant to expand his experimental studies. Eventually he set up what is known today, at least in the English-speaking scientific community, as Boyle's law.[1] It established the inverse proportion of pressure and volume of gases at constant temperature. Boyle is credited with pioneering experimentation as the cornerstone of modern science and with founding the modern science of chemistry. In this, his *The Sceptical Chymist* was his magnum opus. Boyle destroyed many postulates that he proved to be false by experiment. He constructed his own theory that hypothesized that corpuscles and their clusters built up matter and that all phenomena can be

---

[1] Known as Mariotte's law in France, and elsewhere often as Boyle-Mariotte's law

explained by these particles colliding in motion. Although he was one of the founders of the Royal Society, he declined the offer of its presidency on account of his deeply religious demeanor which forbade him from taking an oath for the office.

Left: Portrait of Sir Christopher Wren, engraving by C. Pye after Sir Godfrey Kneller (Wellcome Collection). *Right:* Oil painting by Ernest Board showing Wren making his first demonstration of a method of introducing drugs into a vein, 1667 (Wellcome Collection)

St Paul's Cathedral—Christopher Wren's most famous memorial

Statue of Sir Christopher Wren by Edward Bowring Stephens, 1874, on the court-side façade of Burlington House and plaque at the Old Court House, Hampton Court Green, KT8 (Spudgun67)

Sir Christopher Wren (1632–1723) is best known as an architect. His most famous structure is the St Paul's Cathedral, but he and his associates designed over 50 other churches and many significant secular buildings still extant. He was instrumental in the rebuilding of London following the Great Fire of 1666. He was also an anatomist, astronomer, mathematician, and physicist. Educated at the University of Oxford, he started his career in science there before he was appointed Professor of Astronomy at Gresham College in London in 1657, and a few years later he was elected Savilian Professor of Astronomy at Oxford in 1661. One of the founders of the Royal Society, he later served as its president (1680–1682). Throughout his long life he played an active role in the Society. In 1669 he was appointed Surveyor of Works to King Charles II in recognition of his involvement in rebuilding London. Wren's activities in science were on the broadest possible scale. To mention a few examples, he studied terrestrial magnetism, performed a direct injection of drugs into the bloodstream of a dog, constructed and improved telescopes and microscopes, made pioneering observations in astronomy, investigated various problems in mechanics and meteorology, studied changes in muscle behavior under varying external conditions, and published numerous observations and innovations in optics and mathematics. Gradually, with increasing devotion to architecture, his endeavors in science diminished. In addition to the blue plaque presented above, there is a memorial plaque at his burial place in St Paul's Cathedral. It advises visitors looking for Wren's memorial just to look around—meaning that the Cathedral *is* his memorial.

  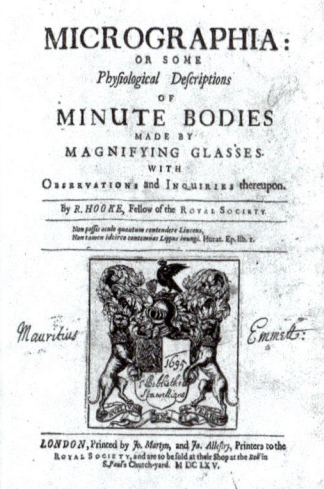

From left to right: Posthumous portrait of Robert Hooke by Rita Greer (Wikimedia); plaque inscribed in the pavement at the Monument, EC3; and the title page of Robert Hooke's *Micrographia* (Wellcome Collection)

Robert Hooke (1635–1703) was born on the Isle of Wight to an impoverished family. He did not have an easy start on his way to his later fame and wealth. From early on, he showed strong curiosity combined with mechanical ingenuity. He could have hoped, at the time, to find prosperity as an artisan or a watchmaker. Instead, he moved to Oxford, became Robert Boyle's assistant, and excelled in helping him with his experiments. He stayed in Boyle's service from 1655 to 1662, crucial years during which he developed a lifelong dedication to science. During this apprenticeship, he conducted his own experiments and discovered the fundamental physical law of elasticity, known as Hooke's law. His connection to the Royal Society happened through Boyle. Building and demonstrating experiments was the essence of the early activities of the Society. Increasingly, experiments were performed during its weekly meetings. The need arose to have someone employed by the Society to be responsible for the experiments. Hooke was the natural choice for curator of the experiments. He proved to be a genius in this role and probably can be counted as the first person in Western history to make his living by being engaged in pure science. He attended the meetings, demonstrated his experiments and the experiments of others, and had become a member of the inner circle of scientists even before he was elected Fellow in 1663.

From this point, Hooke's career was on the ascent. He was appointed Professor of Geometry at Gresham College and kept this position till the end of his life. Boyle, Wren, and Hooke were the most prominent Fellows in the early period of the Society. Then, of course, Isaac Newton came. Hooke was a polymath and produced discoveries and innovations in a great variety of scientific disciplines; suffice it to read the competencies inscribed on his plaque displayed above: "natural philosopher; horologist; astronomer; microscopist; geologist; physiologist; architect"; in short, "England's Leonardo."

Hooke had strong devotion to the Royal Society. He published his *Micrographia* in 1664 in which he summarized his work in experimental microscopy. He proudly added "Fellow of the Royal Society" to his name on the title page. Hooke's achievements in architecture and as Surveyor to the City of London flourished following the Great Fire of London in 1666. In addition to the plaque at the Monument and one on St Helen's Church at Undershaft, his first burial place, he has memorial plaques at Westminster Abbey (installed in 2005) and at St Paul's Cathedral. He had a difficult personality and kept grudges, but his background and the conditions of his early life, so different from his peers, may have been responsible for his irritability. He accused Newton of plagiarism, which offense Newton never forgave during Hooke's life or even after he had died. During the past 300 years, Hooke may have not been accorded the recognition he deserved because he labored in the shadow of Christopher Wren and Isaac Newton. This appears to be changing lately. Hooke's career may also be an example how science, perhaps more than any other human endeavor, may open vistas of elevation in society, overcoming barriers of disadvantageous birth and poverty—and even flaws of character.

Sir Isaac Newton, line engraving after J. Vanderbank, 1720, and the title page of Newton's *Principia* (1687, first edition) (both, Wellcome Collection)

Sir Isaac Newton (1643–1727) has been called "the greatest name in the history of the Society."[2] His memorials are accounted for later in this chapter. Here we mentioned him as President of the Royal Society. In fact, his presidency solidified the activities and prestige of the Society. He was elected Fellow in 1672 in recognition of his invention of the reflecting telescope. His animosity with Hooke dated from this very act as Hooke questioned the novelty of Newton's invention. This animosity intensified upon Newton's publication of his *Principia*. Hooke's belligerence was not dampened by Newton's famous statement, "If I have seen further, it is by standing on the shoulders of giants." Hooke, who was of small stature, took the possible compliment as an affront. Newton was president from 1703, the year Hooke died, until his own death. The Royal Society, whose birth signifies the birth of modern science, has flourished ever since.

---

[2] Quoted by Adrian Tinniswood, *The Royal Society and the invention of modern science* (New York: Basic Books, 2019), p. 75, from E. N. da C. Andrade, *A Brief History of the Royal Society* (London, 1960), p. 6

Bust of Dame Miriam Rothschild by Marcus Cornish, 2004, in the lobby of the Royal Society

Dame Miriam Rothschild (1908–2005) was an entomologist held in high regard by her colleagues in her field. They referred to her as "entomologist extraordinaire," "Queen Bee," or "Lady Flea." She was educated at home and never entered university. Even though she published 350 papers, she did not consider herself a scientist; rather, she regarded herself as the last of the old-time naturalists, a leftover from the nineteenth century. When we visited her in 2002, she lamented that the general knowledge of scientists is getting narrower and narrower, and thus they can no longer find a common language with their colleagues in related fields. She remarked, "today you cannot just sit down and talk about insects in general. All you can do is talk about the hind leg of a bee."[3] The Royal Society elected her a Fellow in 1985. She received honorary doctorates from eight universities, among them Oxford and Cambridge, and she was knighted in 1999.

---

[3] Magdolna Hargittai, *Women Scientists: Reflections, Challenges, and Breaking Boundaries* (New York: Oxford University Press, 2015), p. 158

Royal Institution

The Royal Institution, Albemarle Street, W1

The Royal Institution of Great Britain was founded in 1799 at the urging of Count Rumford, who proposed an "Institution for Diffusing Knowledge" in London. In the same year the group bought a building in Albemarle Street and Rumford became the first secretary of the Institution. The goal of the Institution has been science education. In more recent years, the Institution has sponsored a Davy-Faraday Laboratory for scientific research in well-defined, selected areas. Currently one of these focal areas is nanoscience. To characterize the level of scholarship of the Royal Institution, 15 scientists associated with it have received a Nobel Prize, 5 in each of the 3 categories: Physics, Chemistry, and Physiology or Medicine. Scientists of the Royal Institution have been involved in the discovery of ten new elements.

From left to right: Sir Benjamin Thompson—Count von Rumford's plaque, 168 Brompton Road, SW3; Caricature "The comforts of a Rumford Stove," 1800, by James Gillray; and Rumford's engraving portrait by T. Müller (both, Wellcome Collection)

Sir Benjamin Thompson, Count Rumford (1753–1814), was born in Massachusetts and as a child in the American colony had a colorful life. He enrolled in his village school and walked ten miles to attend lectures at Harvard College. He did not have much hope for a career, but his fortunes changed when he married a wealthy and influential widow in Rumford, what is today Concord, New Hampshire. He joined the British military and helped the British in the American Revolutionary War. Later he abandoned his wife. Her death made it possible for him to marry the widow of the French chemist Antoine Lavoisier.

Thompson was interested in science from childhood and during the Revolutionary War conducted experiments with gunpowder, which made him known when he moved to London in 1781. An 11-year sojourn in Bavaria followed with heightened and diverse activities and the receipt of the title Count of Rumford. Of his scientific experiments, those in thermodynamics stood out, and he utilized his experience in industrial applications, in particular in his designs of industrial furnaces.[4] Other areas included studies of light and photometry. Back in London, he was an active participant of the city's societal and scientific life. He and Joseph Banks (see later in this chapter) established the Royal Institution in 1799. When Rumford married his second wife in 1804, they moved to Paris. Although they separated after 3 years, he stayed in Paris and continued his research there. His grave is in the Auteuil cemetery in Paris.

---

[4] To appreciate the importance of furnaces in science, suffice it to remember the motivation of Henri Moissan's Nobel Prize in Chemistry in 1906: "in recognition of the great services rendered by him in his investigation and isolation of the element fluorine, and for the *adoption in the service of science of the electric furnace called after him*" (our emphasis). Moissan received his award in competition with Dmitri I. Mendeleev, who was a strong nominee in the same year for his Periodic Table of the Elements.

From left to right: Bust of Sir Humphry Davy at the Royal Institution; his engraved portrait by J. Thomson after a painting by H. Howard (Wellcome Collection); and his statue by M. Noble at Burlington House

Sir Humphry Davy (1778–1829) was both a chemist and a poet. He started from very modest circumstances, but before he died, he had become the first scientist ever elevated to baronetcy and had been a President of the Royal Society. He did not distinguish himself in his schooling in Penzance and in Truro, both in Cornwall, but his course of study was not sufficiently challenging. He recognized early that he himself would have to watch out for his own education. After apprenticing to an apothecary at the age of 17, he soon became a chemist and started experimenting. His scientific interests were piqued by various practical problems, such as the corrosion of metals. This approach of starting from a practical problem remained characteristic for his entire career. For some time, he worked at the Pneumatic Institution in Bristol where a great variety of gases were investigated for possible utilization. Davy suggested that the "laughing gas," dinitrogen oxide or nitrous oxide, N_2O, might be applied in surgical operations. This was a precocious inference as anesthetics would be administered during surgery only half a century later.

Davy started publishing in 1799 on heat, light, and the properties of gases. In 1801 Count Rumford, Joseph Banks, and Henry Cavendish interviewed him at the Royal Institution and offered him a position. Soon enough Davy began lecturing at the Institution. Even when his lectures were about specific scientific questions, they extended to general observations on the role of scientific discovery in the progress of civilization. He demonstrated genuine showmanship in his lectures and peppered them with poetry and intriguing, sometimes dangerous, experiments. Consequently, his talks became immensely popular. Davy invented electrochemistry and discovered a large number of chemical elements. He invented the Davy lamp trying to improve the safety of coal mines. He excelled in theoretical considerations. His definitions for acids and bases stayed generally accepted for a hundred years.

By inhaling nitrous oxide, he may have injured his own health, for the reaction of nitrous oxide with water in the mucous membranes forms strong nitric acid. In 1812, he had another laboratory accident while working with nitrogen trichloride. At least in part, this mishap prompted Davy to hire Michael Faraday to help him with transcription. Faraday assisted Davy in the laboratory and, at least at the beginning, served as his valet. Faraday accompanied Davy on his European tours. They traveled to France where Davy received a high award from Napoleon although he was disappointed it was not handed to him personally by the Emperor. They continued to Italy where Davy demonstrated his experiments and on to Austria and Turkey. Davy's contributions to science and its popularization were recognized with memberships in the most prestigious British and international learned societies. Davy died during a stay in Switzerland and was interred in Geneva. There is a marble tablet as his memorial on the wall of the Chapel of St Andrew at Westminster Abbey.

Left: Michael Faraday's marble statue (1874) at the Royal Institution; John Foley was its sculptor who could not complete it on account of his death. Foley's understudy, Thomas Brock, completed it. Right: bronze relief of Michael Faraday's Christmas lecture, 1856, as a door decoration of the former ICI building, 9 Millbank, SW1

A bronze copy of the original Foley marble Faraday statue in front of the Institution of Engineering and Technology, 2 Savoy Place, WC2R, with the induction ring highlighted

Bust and medal of Michael Faraday at the Royal Institution

Michael Faraday (1791–1867), one of the foremost scientists of the world, like Hooke and Davy, had a difficult beginning. He came from a poor family and after the most basic schooling he had to educate himself. From age 14, for 7 years he apprenticed with a bookbinder and bookseller. This long apprenticeship allowed him to read voraciously; through self-study he became interested in science. In 1812, he received tickets to Humphry Davy's lectures at the Royal Institution and their meeting had a fortunate outcome. Faraday became an assistant—sometime valet—to Davy. He accompanied Davy on a 2-year European tour, which gave Faraday an excellent opportunity to meet foreign scientists. In research, he first collaborated with Davy in organic chemistry; eventually, he pursued his own projects, for example, he discovered benzene independently. He worked in a variety of areas in physics and chemistry. He made discoveries in electrochemistry and described what have become known as Faraday's laws of electrolysis. Some of his results belong to an area that is called nanoscience today. His most seminal achievements were in electricity and magnetism. He was the first to arrive at the concept of the electromagnetic field, which led to the principle of electromagnetic induction and his inventions of electromagnetic devices, which formed the foundation of electric motor technology. It is largely due to Faraday that electricity has become a mainstay of our everyday lives, not merely a phenomenon studied by physicists. Faraday was not versed in mathematics; thus, he could not communicate in the language of mathematics. Instead, he had excellent abilities to convey complex scientific concepts in a simple, easy-to-perceive language that a broad audience could absorb. His lectures were immensely popular. It remained to James Clerk Maxwell to express the interrelationship of electricity and magnetism in the language of mathematics, hence the famous Maxwell equations (see below).

Plaques of Michael Faraday: (left) at 48 Blandford Street, W1U, the location where he apprenticed, and (right) at the Larcom Street entrance to the Walworth Clinic, SE17 (this image, Matt Brown)

Michael Faraday Memorial on Elephant Square, SE1, at the Elephant and Castle Metro Station

In later years, Faraday was careful in choosing which honors to accept and which to decline. He wanted to stay a simple man for moral and religious reasons. He declined knighthood, but accepted Fellowship of the Royal Society, though not its presidency, which was offered to him not once, but twice. He accepted honorary memberships in foreign academies of the United States, France, Sweden, and the Netherlands. With Prince Albert's strong support, Faraday was given a house in London, free of all expenses and upkeep. Well in advance, he declined to be buried at Westminster Abbey; rather, he is buried at Highgate Cemetery. Nonetheless Westminster Abbey installed a memorial stone for him in its nave, near Isaac Newton's tomb. It was unveiled in 1931 and replaced by a plaque made of metal in 1976.

Bust of Mary Somerville at the Royal Institution and her lithographic portrait after J. Phillips (Welcome Collection)

Mary Somerville (née Fairfax, 1780–1872) and Caroline Herschel (Chap. 2) were jointly the first female honorary members of the Royal Astronomical Society in 1835. Somerville was called "the queen of science" of her time. She grew up in Scotland and had rather accidental schooling at home, in village schools, in an academy for girls in Edinburgh, and from individuals whom she encountered. Her thirst for knowledge gradually led her not only to become versed in a variety of fields, but also to embark on independent research, in particular in mathematics and astronomy. Skill at writing seemed to come naturally to her, and she further refined her style by studying the writings of Euclid, Newton, Laplace, and others. Her first recognition came for having solved some mathematical problems. Her second husband, Dr. William Somerville, encouraged her to continue her studies and investigations in the physical sciences. Because he had an important medical appointment, the couple moved among the best society of scholars and artists in London, such as Walter Scott, Charles Babbage, and J. M. W. Turner, and, eventually, Ada Lovelace. Somerville studied magnetism and light and kept publishing her findings. When she was asked to translate Laplace's monumental treatise, she did something novel. Rather than merely translating it from the French original into English, she translated what was presented in the language of algebra into a common language. The title was *The Mechanism of Heavens* (1830). It was an immediate success and remained a textbook in Cambridge for 50 years. Her second book, *On the Connexion of the Physical Sciences* (1834), was an even greater success. Then her *Physical Geography* (1848) became the textbook of choice in English for more than half a century. Somerville became a great disseminator of science; her fame in this regard peaked with her fourth book, *Molecular and Microscopic Science* (1869), which would be an intriguing title even today. Her autobiography, *Personal Recollections* (1874), based on old-age reminiscences and correspondence, appeared posthumously.

John Tyndall relief at the Royal Institution and Visit of Queen Victoria to the Loan Collection of Scientific Apparatus at South Kensington; Professor Tyndall demonstrating the action of the foghorn, wood engraving by T. B. Wirgman (Wellcome Collection)

John Tyndall (1820–1893) grew up in southeast Ireland, worked for the fledgling railway industry, and in 1847 decided to continue his education. He became acquainted with modern experimental physics at the University of Marburg in Germany with Robert Bunsen and other leading professors. Tyndall's research of magnetism brought him fame and recognition. He conducted his most extensive investigations of the air—the atmosphere of the Earth, studying the ability of various components of the air to absorb heat. His work still has relevance to our current worries about global warming.[5]

Another focus of Tyndall's research was marine navigation, which he wanted to make safer by devising methods to warn ships away from hazards despite fog. He studied the acoustics of the atmosphere. His conclusion was that a loud, low-pitched sound would be most effective in order to minimize the scattering of sound by the fog. The device was called the foghorn and became a boon to marine navigation; it even elicited Queen Victoria's curiosity. The recognition of the Tyndall effect was another output from his research of the atmosphere. It is manifested by the scattering of light when a light beam passes through a gas (or a liquid), which contains finely dispersed particles. The demonstration of the Tyndall effect is often part of the school curriculum even today. An enthusiastic mountaineer, Tyndall's field observations also enriched the science of glaciers and their motion.

Tyndall was Professor of Physics at the Royal Institution between 1853 and 1887 where he was a popular lecturer and a prolific author of books about science. His lecturing and writing added considerably to his income, much of which he donated to benevolent causes. He was dedicated to the separation of science and religion and supported Darwin's theory of evolution. He was tireless in disseminating his views, whether as a private researcher or as president (1874) of the British Association for the Advancement of Science. He was a great advocate of the unity of "two cultures"—arts and sciences—as witnessed by the following quote from an address he gave before the Association: "The world embraces not only a Newton, but a Shakespeare—not only a Boyle, but a Raphael—not only a Kant, but a Beethoven—not only a Darwin, but a Carlyle. Not in each of these, but in all, is human nature whole. They are not opposed, but supplementary—not mutually exclusive, but reconcilable."[6]

---

[5] What we call greenhouse effect is when gases in the atmosphere absorb the heat from the Sun. While this is what makes the Earth habitable, lately, the increased production of the so-called greenhouse gases, such as carbon dioxide, has caused the atmosphere to trap an increasing amount of heat and the Earth to warm up. The continuation of this process would have catastrophic consequences.

[6] John Tyndall, Address delivered before the British Association assembled in Belfast, 1874. See in R. Schiller, *Between One Culture: Essays on Science and Art* (Springer Nature Switzerland, 2020), p. 144.

The Royal Institution, which embodies Tyndall's legacy, has been famous for its activities in the dissemination of knowledge and in particular for its Christmas Lectures initiated by Michael Faraday. These Lectures have been running since 1825 with the exception of 1939–1942. From 1966, the Christmas Lectures have been televised. Faraday personally held 19 of these Lectures and he was the presenter in every year during the period 1851–1860. Today, the Lectures remain open to the public; they are held to a selected single topic; and the presenters take special care to make their presentations accessible for a general audience. Beside the special Christmas Lectures, there are the Friday Evening Discourses presented by equally distinguished scholars. The Institution opened the Faraday Museum in its main building, which houses, among others, Faraday's original laboratory, the site of his research in the 1850s. These lectures are a reminder that the habits of mind fostered by Faraday and Tyndall continue to inspire new generations of scientists and laypersons.

Sir James Dewar. Left: Painting by H. Jamyn Brooks of Dewar giving a Friday Evening Discourse in 1904 on liquid hydrogen. Right: Plaque by Sir Bertram Mackennal, 1925

Sir James Dewar (1842–1923) was an experimental chemist whose name is well known among experimentalists due to his invention of the Dewar flask: a device for insulating a storage vessel to keep the contents at a higher or lower temperature than its surroundings. His invention, even if not his name, is also well known among the general population, because the thermos (which is a trademark) is a Dewar flask. Also known for his success, in 1898, at liquefying hydrogen, he held a number of Christmas Lectures on a great variety of topics, such as the soap bubble, atoms, alchemy, meteorites, light and photography, clouds, frost and fire, and gases and liquids. In 1912, he gave one more with the title "Christmas Lecture Epilogues." He held high offices in scientific organizations and was highly decorated by British and international scientific institutions.

Sir William Lawrence Bragg giving the 1961 Christmas Lecture on electricity and his photograph by and courtesy of the late David Shoenberg

Sir William Lawrence Bragg (1890–1971) had a brilliant career even before his association with the Royal Institution (1954–1971). He studied at the University of Adelaide in Australia, where his father, Sir William Henry Bragg (1862–1942), was Professor of Mathematics and Physics. They moved to England in 1908 where William Lawrence completed his studies in physics at the University of Cambridge. When Max von Laue and his two associates in Munich showed in 1912 the wave nature of X-rays (for which Laue received the Nobel Prize in Physics in 1914), the two Braggs immediately recognized the relationship between the wavelength of X-rays and the interatomic distances in crystals. Thus X-ray crystallography was born, one of the pivotal moments in the progression of twentieth-century science. The two Braggs were jointly awarded the Nobel Prize in Physics in 1915, when William Lawrence was 25 years old, to this date, the youngest ever science laureate.

Following defense-related work in World War I, W. L. Bragg was Professor of Physics at Manchester University, then, briefly, Director of the National Physical Laboratory. In 1938 he was appointed Cavendish Professor and in charge of the Cavendish Laboratory—as Ernest Rutherford's successor—in Cambridge. During World War II, he was an advisor in defense-related work. After the war, it was a big question what the future of the Cavendish Laboratory would be. It used to be a world leader in nuclear physics under Rutherford, but this was no longer the case. In a bold move, Bragg decided to change the main thrust of the Laboratory and encouraged the development of two new areas in which the laboratory had no track record: radio astronomy and molecular biology. Molecular biology was an unorthodox focus for any major laboratory. Both choices proved to be seminal decisions for science history, and both led to resounding success (and Nobel Prizes).

From 1954 until the end of his life, Bragg lived at the Royal Institution and held a variety of functions in succession. He revitalized the Institution, including the Christmas Lectures and the Friday Evening Discourses. He created a circle of sponsors about the Institution and placed it on a solid financial foundation. The Davy-Faraday Laboratory within the Institution reached important results in protein crystallography. Bragg's encouragement of new ideas and research directions to this date have produced seminal discoveries and Nobel Prizes in molecular biology: the latest in 2017 and 2018. When Bragg moved from Cambridge to the Royal Institution, he missed gardening, which he used to enjoy. So he advertised himself for gardening work and was hired as an ordinary gardener for a few hours weekly. This pursuit continued until a visiting friend of his employer asked her about what Sir Lawrence was doing in her garden.

Left: Painting of Lord Porter by an unknown artist. Right: The cover of the magazine *The Chemical Intelligencer* (Springer, Vol. 4/4, October 1998) showing Queen Elizabeth II with Lord Porter during her visit in 1973 at the Royal Institution

George Porter (Lord Porter, 1920–2002) was Director of the Royal Institution for 20 years (1966–1986). He was a Nobel laureate chemist whose World War II experience with radar was crucial in developing his unique experiments for investigating extremely fast chemical reactions. His activities in popularizing science gave him fame. During a televised Christmas Lecture, he cleverly illustrated the law of inertia. He presented a beautifully laid table setting, and after repeating three times "I believe in Isaac Newton," he yanked out the tablecloth to leave the china and crystal intact on the table. On the following New Year day, Lord Porter and a few others were the Queen's guests for a luncheon with the participation of the Royal Family. The party was walking toward lunch in a drawing room with its large table set for 12 people and loaded with silver and gold plate. The young Princess Anne urged Lord Porter: "I liked your demonstrations, especially the one where you cleared the table. Go on, try it now!"[7]

---

[7] Of course, Lord Porter did not. This story was narrated to us by Lord Porter on September 11, 1997, in his office at Imperial College (I. Hargittai, *Candid Science: Conversations with Famous Chemists*, edited by Magdolna Hargittai, London: Imperial College Press, 2000; Chapter 38, "George Porter," pp. 476–487; the story is described on p. 480).

Physicists, Astronomers, and Mathematicians

Left: Pythagoras among poets in the frieze on the southern side of the Albert Memorial, surrounded by Virgil, Dante, and Homer, by Henry Hugh Armstead. Right: Statue of Archimedes by William Frederick Woodington on the central balustrade of Burlington House

Pythagoras (c.570–c.495) was an ancient Greek philosopher and mathematician whose teachings have extended over millennia. He figures here because of his mathematics, including what is known popularly as the Pythagorean theorem. It establishes the relationship among the three sides of a right triangle: the sum of the areas of the two squares of the legs (a and b) equals the area of the square of the hypotenuse (c), or as expressed by an equation, $a^2 + b^2 = c^2$.

Archimedes (c.287–c.212 BCA) was a Greek polymath—mathematician, physicist, inventor, and astronomer—one of the most important scientists of antiquity. He could figure equally among the inventors just as well as among the scientists.

From left to right: Statue of Galileo Galilei by Edward William Wyon on the balustrade of the East Wing, Burlington House; portrait by J. Sustermans, nineteenth/twentieth century, in the Uffizi, Florence (Wellcome Collection); and bust of pear wood by Giovanni Battista Foggini, between 1670 and 1710, in the Queen's House, Greenwich Park, Romney Road, SE10 (this image, Matt Brown)

Galileo Galilei (1564–1642) was a physicist, astronomer, inventor, and polymath from Pisa. He is a legendary figure from the dawn of modern science. He built instruments that were necessary for making progress in physics and astronomy and contributed a great deal to both pure science and engineering. Robert Hooke's meeting with him stimulated Hooke in becoming a lifelong, dedicated scientist.

Left: William Blake's "The Ancient of Days," 1805 (Wikimedia). Right: *Newton* after William Blake by Eduardo Paolozzi, 1995, British Library forecourt, 96 Euston Road, NW1

The artist William Blake (1757–1827) resented the Enlightenment. His painting reproduced above supposedly says that in its blind endeavor, Science tends to ignore beauty. However, one need not take Blake's interpretation as gospel, for both his painting and Paolozzi's *Newton* radiate beauty. The sculpture is a profound introduction to the British Library.

From left to right: Statue of Isaac Newton by Joseph Durham above the portico of Burlington House; his bust by William Calder Marshall, 1874, on Leicester Square, WC2 (we took this photograph in 2003; the bust disappeared during the reconstruction of Leicester Square in 2010–2012); and Newton's plaque, 87 Jermyn Street, SW1Y

(Left) Statue of Newton on the façade of the former building of the City of London School and (right) one of the seven relief panels by William Grinsell Nicholl on the façade of the Oxford and Cambridge Club, 71–77 Pall Mall, SW1Y. It depicts Newton holding a globe and explaining a point to a group of adults and children

Sir Isaac Newton has already been introduced as the longtime president of the Royal Society and the greatest name among all its Fellows in its history. His father was a simple man, a farmer, who died a few months before Isaac was born. After his mother remarried, Isaac was brought up by his maternal grandparents, whom he did not like. The teenaged Newton has been described as a meditative, shy, and slow boy, who was bullied at school. One day, however, he turned violently against the school bully. This may have been a turning point for the better in his development, both physically and academically. Nonetheless, his fate appeared to be charted as a farmer, except that during his school years he boarded with the village druggist whose books Newton began reading voraciously. Furthermore, he developed an interest in devising complex mechanisms, such as water clocks, sundials, and flying kites. He was fascinated by sunshine and the winds and recognized that he must learn math to understand them. He was lucky to have the possibility to attend college at Cambridge, first as a law student, then drawn by the works of Descartes, as a student of math and physics. When a plague outbreak forced him to leave Cambridge for home, he experimented with prisms and learned more about sunshine. His course for science was set. He graduated with an A.D. degree from Trinity College at the age of 26—not early by any consideration. However, early on, he became interested in gravitation and made much use of what he had learned about the relevant discoveries of Kepler and Galileo. Newton's discovery of the universal law of gravitation manifested his tremendous insight into how Nature operates. Nor did his mechanical ability languish. Recognizing the need for and importance of instruments in research, he created the first reflecting telescope. In short order, he made seminal discoveries in mathematics, optics, and astronomy. He invented calculus, which he called the "method of fluxions." The German Gottfried Wilhelm von Leibniz also invented calculus and he called it the "method of differences." A controversy of many years ensued with Newton's priority finally recognized. His *Principia*—formally known as *Philosophiae Naturalis Principia Mathematica*—of 1687 is one of the milestone publications in modern science. The lateness and interruptions of his university work did not compromise his career, for he was appointed to the Lucasian Chair of Mathematics at Cambridge very early, in 1669. From his many disputes and other actions, the impression of an imperial personality emerges. Anecdotes abound about his absolute devotion to science, without paying much attention to anything else. Still there seem to have been exceptions. His academic activities did not make him rich, whereas becoming Master of the Royal Mint proved to be a lucrative occupation. He had a weakness in his infatuation with alchemy, and he kept attempting to turn mercury into gold. Shortly before his death, he is known to have made this widely quoted statement: "I know not what I may appear to the world, but to myself I seem to have been only like a boy playing on the sea-shore, and diverting myself in now and then finding a smoother pebble or a prettier shell than ordinary, whilst the great ocean of truth lay all undiscovered before me."[8] By any other estimate, Newton was one of the greatest scientists of the world, if not the greatest, a distinction he shares with Albert Einstein. Newton lived near Leicester Square in downtown London. For a long time, Newton's bust stood on Leicester Square, but disappeared during the recent renovation of the Square.

---

[8] David Brewster, Ed., *Memoirs of the Life, Writings, and Discoveries of Isaac Newton*. Vol. II (Edinburgh: Thomas Constable and Co., 1855), Chapter 27, as quoted in Alan L. Mackay, *A Dictionary of Scientific Quotations* (Bristol: Adam Hilger, 1991), p. 181.

The former building of St Olave's Grammar School, Queen Elizabeth Street, SE1, and Isaac Newton's portrait carved into stone on the front of the building

St Olave's Grammar School was established in 1571 at Queen Elizabeth and Tooley Streets; its main building was designed by the architect E. W. Mountford, 1894. It has three semicircular pediments. The one on the left depicts Newton, flanked by two figures symbolizing Science and Philosophy. The School itself moved to Orpington in 1968.

Statue of Gottfried Wilhelm Leibniz in a ground story niche of the East Wing of Burlington House and his engraved portrait (Wellcome Collection)

Gottfried Wilhelm Leibniz (sometimes Leibnitz, 1646–1716) was a German polymath. His most notable accomplishment in mathematics was the development of calculus, the differential and integral calculation of continuous change. Isaac Newton developed calculus at about the same time, independently. Other areas of Leibniz's activities included philosophy, physics, and technology.

Statue of Pierre-Simon Laplace by Edward William Wyon on the balustrade of the East Wing, 6 Burlington Gardens, W1, and an etching portrait by Charles-Nicolas Cochin, 1762 (Wellcome Collection)

Pierre-Simon Laplace (La Place, 1749–1827) was much more than a mathematician. He made important contributions in engineering, physics, astronomy, and philosophy as well. He has been called the French Newton. His schooling started in the village of his birthplace and continued at the University of Caen, where he majored in theology. He wrote his first science paper while yet a student and published it in a fledgling journal initiated by Joseph-Louis Lagrange, another great in science. At the recommendation of Jean le Rond d'Alembert, one of the French Encyclopedists, Laplace obtained a position in the prestigious École Militaire, where he embarked on his research career. Following an unsuccessful attempt to elect Laplace to the French Academy of Sciences in 1771, he was elected in 1773. He was only 24. In 1780–1784, Laplace and Antoine Lavoisier did joint research on the theory of molecular motion. Laplace accumulated milestone discoveries in astronomy, in the theory of tides, and in various applications of his mathematical tools to solving physical problems, and he wrote fundamental books of lasting value.

Laplace's life coincided with turbulent eras in French history, including the Bourbons, the French Revolution, Napoleon and his wars, and the Bourbon Restoration. Although Laplace was always wary of any involvement in politics, in 1899 he served as Napoleon's Minister of the Interior. He was dismissed as there was no need to have a name of great prestige in the government of Napoleon's new order. After Napoleon's fall, Laplace offered his services to the Bourbons. As for his views on religion, there is a famous story. According to one of its versions, Napoleon asked Laplace why he did not mention God in his book about the universe. Laplace's answer was that he had no need of that hypothesis.

Left: Statue of Johann Wolfgang von Goethe in between John Milton and Friedrich Schiller on the southern side of the Frieze of Parnassus at the Albert Memorial. Right: Goethe's portrait, line engraving by Carl August Schwerdgeburth, 1832 (Wellcome Collection)

Johann Wolfgang von Goethe (1749–1832) was a great German writer, a statesman, and a scientist. He wrote that he did not take pride in what he had done as a poet but was proud of his achievements in the science of colors. The general notion is that discoveries in science hardly stay in the collective memory because new discoveries make the previous ones appear obsolete as science continues building up vertically. By this logic, if one scientist does not make a discovery, somebody else will, sooner or later. In contrast, if Goethe had not written *Faust*, nobody else would have, and *Faust* would have stayed unwritten forever. Nonetheless, Goethe considered his most important work what he summarized in his *Theory of Colours*. He did science also in morphology, botany, mineralogy, and geology.

Thomas Young memorial plaque, 48 Welbeck Street, W1, and his portrait, engraving by H. Adlard after Sir Thomas Lawrence (Wellcome Collection)

Thomas Young (1773–1829) has been described as the last person who knew everything. Even though this is not a unique characterization, it says a good deal about the breadth of his interests and expertise. He was a polymath and also a physician. He studied medicine at St Bartholomew's Hospital, the University of Edinburgh, the University of Göttingen, and at Cambridge. In 1799 he opened his private practice at 48 Welbeck Street. Simultaneously with practicing medicine he was a prolific author of scientific papers, but he published them anonymously to avoid hurting his medical practice. In 1801 he was appointed Professor of Natural Philosophy—today the closest title would be Professor of Physics—at the Royal Institution. Here, he was often elected to leadership of various committees dealing with precise time measurement, the determination of longitude, the publication of the *Nautical Almanac*, and suchlike. Among his scientific contributions, he introduced the "Young's modulus" to give a qualitative characterization of elasticity. He discovered astigmatism; and he did research on the wave nature of light. He identified a good number of Egyptian hieroglyphics, but his work in this area remained incomplete. His profile medallion on a memorial tablet by Sir Francis Chantrey was unveiled in 1834 in the Chapel of St Andrew at Westminster Abbey, but he is buried elsewhere.

William Thomson, Lord Kelvin, photograph by W. & D. Downey (Wellcome Collection) and his plaque, 15 Eaton Place, SW1

William Thomson—Lord Kelvin (1824–1907) was a mathematical physicist and an engineer-inventor (see also Chap. 5). He studied in Belfast, Glasgow, and Cambridge. Already at the age of 17, he published substantial papers on thermodynamics and electricity. He was appointed Professor of Natural Philosophy at Glasgow University in 1846 and he stayed in this position until his retirement. While in Glasgow, he formulated the First and the Second Laws of Thermodynamics and further developed his mathematical analysis of electricity. He had a fruitful interaction by correspondence with another physicist, James Prescott Joule (1818–1889). Thomson's theoretical and interpretational work and Joule's experimental research eminently complemented each other. Thomson recognized that heat loss was loss to man but not to the material world. He speculated about the heat death of the universe. He determined the correct value of the absolute zero temperature (–273.15 degrees Celsius and –459.67 degrees Fahrenheit). The absolute temperature scale is named after Kelvin; the absolute temperature is expressed in kelvin ("degree" does not figure in it). He did crucial engineering work in connection with the transatlantic cable, improved the mariner's compass, and was an early user of the long-distance telegraph. In 1874 he proposed by telegraph and received his future second wife's "yes" by telegraph. Queen Victoria knighted Thomson in 1866 for his contribution to the transatlantic cable. In 1892, the ennobled Sir William Thomson—Lord Kelvin was the first British scientist to join the House of Lords for scientific achievements. There was, though, a political component of this recognition, an acknowledgment of his opposition to Irish Home Rule. He was one of the inaugural members of the most distinguished Order of Merit in 1902. Lord Kelvin was buried in Westminster Abbey. His grave is in the nave, close to the graves of Sir Isaac Newton, Sir John Herschel, and Charles Darwin.

James Clerk Maxwell plaques, at 16 Palace Gardens Terrace, Kensington, W8 (left), and at the Strand Campus of King's College (right)

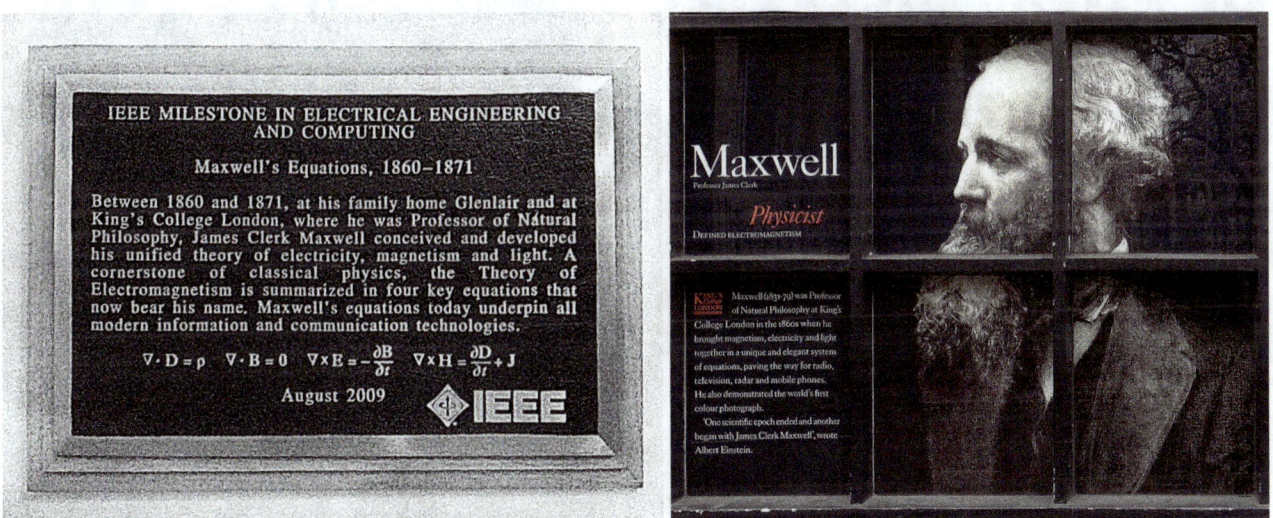

The Maxwell equations on a memorial plaque at King's College (Wikimedia) and a photograph of James Clerk Maxwell in a display at King's College on the Strand Campus

James Clerk Maxwell (1831–1879) united the classical theories of electricity, magnetism, and light in a set of four equations, called the Maxwell equations. Following some home education, when he was 10 years old, he was enrolled in the Edinburgh Academy. He was fascinated by geometry and re-discovered the regular polyhedra on his own. He wrote his first scientific paper when he was 14. In 1847, he began attending the University of Edinburgh. He had two more scientific papers at the age of 18: one in geometry and the other on the physical properties of solids. He was a student at Cambridge, 1850–1856. First he focused on mathematics, but gradually his attention was turning to physics. One of his first papers was on color. This was in 1855, and by then he was allowed to deliver his own presentation; before, others had to do it for him on account of his young age. In 1856, Maxwell returned to Scotland as a professor at Marischal College (today, the University of Aberdeen). One of his projects was a treatise about the rings of Saturn. In the 1980s, the data collected by the spaceship Voyager confirmed his conclusion that the rings consisted of particles. In

1860, Maxwell was appointed to the Chair of Natural Philosophy at King's College, London. The next 5 years were his most productive period when he focused on electricity and magnetism and participated actively in the scientific life of London. In 1865, he resigned from King's College and returned to his birthplace in Scotland. His production of significant treatises continued for the rest of his life, such as *Matter and Motion* in 1876. He published textbooks as well, such as *Theory of Heat* in 1871. That same year, he was appointed to be the first Cavendish Professor of Physics at Cambridge, where he was in charge of developing the Cavendish Laboratory. Maxwell died of cancer in 1879. One of the greatest physicists of all time lived only 48 years. He was buried in the churchyard at Parton, Kirkcudbright, Scotland, close to where he grew up. There was a memorial stone placed in his honor in the nave of Westminster Abbey in 1931, which was replaced by one in iron cast in 1976. Albert Einstein so appreciated Maxwell's contributions that he declared that he stood on Maxwell's shoulders (rather than Newton's).

Plaques of Karl Pearson, 6 Well Road, NW3, and Sir Ronald Aylmer Fisher, Inverforth House, North End Way, NW3 (both, Spudgun67)

Two plaques remember two scientists who helped develop the statistical techniques much used in modern research. Their interests combined statistics and biology, genetics, and other aspects of the so-called life sciences. Karl Pearson (1857–1936) studied mathematics at UCL and at Cambridge, followed by studies in Germany. His interests included history, religion, and social sciences. He taught mathematics at King's College, UCL, and Gresham College, where his involvement in biometry and evolutionary theory developed. He met Francis Galton and fell under his influence. After Galton's death, Pearson spent years in examining Galton's legacy and produced his three-volume biography. Pearson was the first holder of the Chair of Eugenics, subsequently named after Galton. The Chair was established from the portion of Galton's estate that he left to UCL. Pearson's views on superior and inferior races, now distasteful, included his opposition to Jewish immigration to Britain. Pearson's long-lasting contribution was in biometrics where he introduced a number of approaches and tools in statistical analysis that are still fundamental, such as standard deviation, correlation and regression coefficients, and many others. He co-founded the journal *Biometrika*, which published its first issue in 1901.

Sir Ronald Aylmer Fisher (1890–1962) had poor eyesight, which contributed to his unique approach to mathematical problems. He thought about them in geometrical terms rather than writing down lengthy derivations. He studied at Cambridge and became involved in population genetics. He helped the renewal of interest in Darwinism due to his innovative approach to a combination of Mendel's genetics and Darwin's teachings on natural selection. Fisher actively supported the eugenics movement. He believed in racial differences, and his views received much exposure when they were contrary to the statement of UNESCO in 1950, which was a moral condemnation of racism. His achievements in developing the science of statistics brought him international fame and prestigious appointments at UCL and at Cambridge.

From left to right: Bust of Bertrand Russell by Marcelle Quinton, 1980, Red Lion Square, WC1; his plaque, 34 Russell Chambers, Bury Place, WC1A (Spudgun67); and a pavement mosaic "Lucidity" depicting Bertrand Russell by Boris Anrep in the Portico of the National Gallery, Trafalgar Square, WC2N

Bertrand Russell (1872–1970) was a philosopher and writer, logician and mathematician, social critic and political activist, and historian. For our project, he had important work that influenced many areas of science and mathematics, research on artificial intelligence, epistemology, set theory, and computer science, among others. He co-authored *Principia Mathematica* with A. N. Whitehead in which they attempted to provide a logical basis of mathematics. In 1950, he received the Nobel Prize in Literature.

Sir Owen Richardson in a display of King's College, Kingsway, WC2

Sir Owen Richardson (1879–1959) studied at Cambridge and did research at the Cavendish Laboratory, where he studied the emission of electrons by heated wires. It had already been observed by Thomas Alva Edison that a hot wire emitted electrons. Richardson determined that the electron emission depended on the temperature: the higher the temperature, the higher the current density of the electrons. This thermally induced flow of charge from a surface is called the thermionic emission. Richardson suggested a mathematical expression describing the exponential relationship between the electron emission and the temperature. He received the Nobel Prize in Physics for his discovery in 1929. It was the 1928 award, but in that year no Prize was given out as the Swedish judges did not find any prize-worthy contribution in physics. Richardson was a professor at Princeton University between 1906 and 1913. The American physicist Clinton Davisson received his PhD degree in 1911 under Richardson's mentorship and then he married Richardson's sister, Charlotte. Davisson had a brilliant career; he and his associate at Bell Labs, Lester Germer, showed experimentally the wave nature of electrons—electron diffraction. For this, Davisson was a co-recipient of the 1937 Nobel Prize in Physics. The other co-recipient was another Briton, George Paget Thomson, for an independent experiment to the same effect. Richardson returned to the United Kingdom in 1914 and joined King's College in London for the rest of his career.

Plaque of Sir Edward Victor Appleton, Strand Campus of King's College, and his portrait in a display of King's College

Sir Edward Victor Appleton (1892–1965), physicist, received his higher education at Bradford College and at Cambridge University. He served as engineer in World War I. Subsequently, he was at the Cavendish Laboratory, King's College, Cambridge University, the Department of Scientific and Industrial Research, and for the last 15 years of his life, at the University of Edinburgh. His research focused on the propagation of electromagnetic waves; his fundamental scientific contributions date from the mid-1920s. He discovered an upper region of the ionosphere, called now the Appleton layer, which reflects electromagnetic waves, whereas the lower region lets them penetrate. His discovery was instrumental for the development of radar and in establishing reliable long-range radio communication. He received the 1947 Nobel Prize in Physics.

Bust of Patrick Blackett by Jacob Epstein at the Blackett Laboratory—Department of Physics of Imperial College, South Kensington Campus, SW7, and Blackett's plaque at 48 Paultons Square, SW3 (Spudgun67)

Patrick Maynard Stuart Blackett (1897–1974) was an experimental physicist who gained experience with Ernest Rutherford in Cambridge and James Franck in Göttingen at the beginning of his research career. He became a specialist in tracing events involving fundamental particles in a cloud chamber. At Rutherford's suggestion, he bombarded nitrogen atoms with alpha particles and observed the appearance of fluorine atoms as a result. Thus, Blackett was the first person who succeeded in transmuting one element into another—the eternal dream of the alchemists. His career included positions at Birkbeck College of UCL, the Victoria University of Manchester, and Imperial College, where he was the chair of the physics department. In 1948, he received the Nobel Prize in Physics. The citation referred to his discoveries in nuclear physics and cosmic radiation and the Wilson cloud chamber that he modified to make his observations possible. The physics department had other luminaries, discussed below. Blackett's predecessor in charge of the physics department was Sir George Paget Thomson (1892–1975), Nobel Prize co-recipient of the 1937 physics award. He was at Imperial College from 1930 to 1952. One of the well-known physicists was another Nobel laureate, the Pakistani-born Mohammad Abdus Salam (1926–1996), who was a co-recipient of the physics award in 1979.

Frieze-reliefs, from left to right, from Hipparchus to Torricelli at Imperial College, on the chemistry side of Imperial College Road. Archimedes on the extreme left is not seen in this image. The Torricelli relief is highlighted on the right

The ten frieze-reliefs on the Imperial College pediment are listed here:

Archimedes, BC 287–212, Greek mathematician, scientist, and inventor.

Hipparchus, second century BCE, Greek astronomer, geographer, and mathematician.

Geber, possibly the Arab alchemist, eighth century, author of books on alchemy and metallurgy.

Roger Bacon, 1214–1294, philosopher of science, an early advocate of the study of nature through empiricism. His published theories may be regarded as the forerunner of Francis Bacon's scientific method. In Roger Bacon's *Opus Majus*—Greater Work—he discussed mathematics, optics, astronomy, and other areas of human knowledge.

Nicolaus Copernicus, 1473–1543, Polish astronomer, mathematician.

William Gilbert, 1544–1603, English physician and natural philosopher.

Galileo Galilei, 1564–1642, Italian astronomer, inventor, and polymath.

Johannes Kepler, 1571–1630, German astronomer, scientist, and mathematician.

René Descartes, 1596–1650, French mathematician, philosopher, and scientist.

Evangelista Torricelli, 1608–1647, Italian physicist and mathematician.

The Department of Physics was a component of the Imperial College Faculty of Natural Sciences, whose predecessor was the former Royal College of Science. Its origin goes back to the Royal College of Chemistry, which figures later in this chapter.

Left: Pavement mosaic "Leisure" by Boris Anrep displaying Einstein's famous equation in the Portico of the National Gallery. Right: Albert Einstein's photograph (US Department of Energy Photography)

Albert Einstein (1879–1955), one of the greatest physicists of all time, paused in Britain in fall 1933 on his way to seek asylum in the United States. He spent some time in the country, had interactions with statesmen Winston Churchill and Lloyd George, and gave a talk at a rally in London on October 3 (Chap. 1).

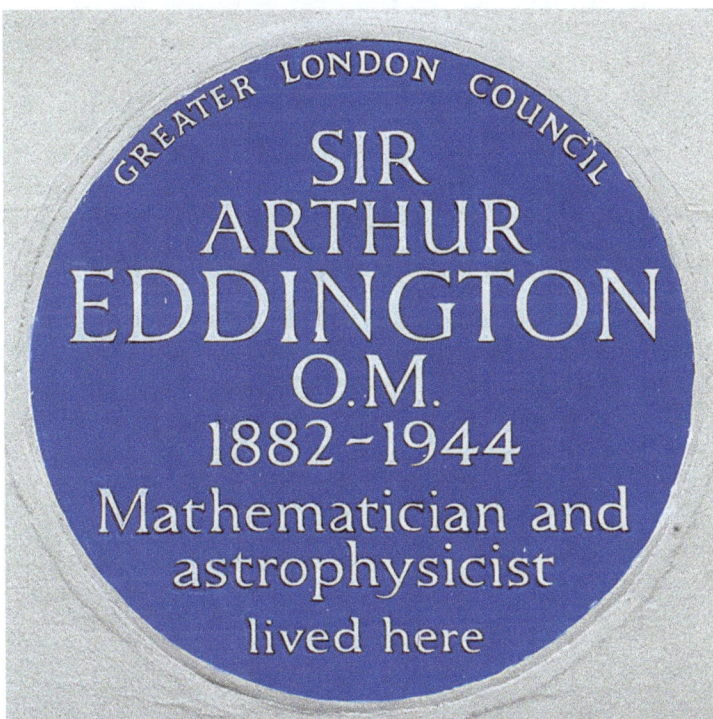 

Plaque of Sir Arthur Eddington, 4 Bennett Park, SE3 (Spudgun67), and his portrait (Smithsonian-Dibner Library)

Sir Arthur Stanley Eddington (1882–1944) was a physicist, astronomer, and a philosopher of science. He received his degrees in physics at what is today the University of Manchester (then, Owens College) and at Cambridge. He started his professional career at the Royal Observatory in Greenwich and continued at the Cambridge Observatory, where he became director. He was an early supporter of Einstein's theory of general relativity. Eddington and the Astronomer Royal Sir Frank Dyson (Chap. 2) conducted expeditions to observe solar eclipses. Their observations, published on May 29, 1919, confirmed Einstein's theory.

Beside this famous contribution, Eddington had other pioneering ideas in astronomy. From today's perspective, his anticipation of thermonuclear energy is rather intriguing. He advanced the idea that the stars are largely composed of hydrogen and the fusion of hydrogen atoms into helium liberates enormous amounts of energy, in keeping with Einstein's equation $E = mc^2$. This was around 1920, that is, decades before these ideas could be proven correct. Eddington was a popular author and lecturer of science and the philosophy of science.

Photograph of Leo Szilard (US Department of Energy Photography) and illustration to the story of where his initial thoughts about nuclear energy were born

Leo Szilard (1898–1964) has only a virtual memorial in London, but it is rather famous. It is the crossing at Southampton Row and Russell Square, WC1. Szilard was one of the five so-called Martians—Jewish-Hungarian physicists who settled in the United States.[9] They contributed greatly to the defense of their chosen new homeland and the Free World, first against Nazi Germany in World War II and later against Stalinist Soviet Union in the Cold War. Szilard spent some time in London in 1933–1934 as part of his transition from Germany to America. During this time, he developed the ideas of nuclear chain reaction and critical mass, filed for patents for the nuclear chain reaction (the basic concept for the atomic bomb), and deposited it with the British Admiralty. In 1939, after nuclear fission had been discovered in Germany, Szilard wanted to warn President Franklin D. Roosevelt of the dangers of a German atomic bomb, so Szilard convinced Albert Einstein to send a letter to that effect to the President. Legend has it that one day back in London as Szilard was waiting for the traffic light at the corner of Southampton Row and Russell Square, he suddenly realized that "if we could find an element which is split by neutrons and which would emit *two* neutrons when it absorbed *one* neutron, such an element, when assembled in sufficiently large mass, could sustain a nuclear chain reaction" (italics in original).[10]

---

[9] The other four were Theodore von Kármán, Eugene P. Wigner, John von Neumann, and Edward Teller. See, e.g., I. Hargittai, *The Martians of Science: Five Physicists Who Changed the Twentieth Century* (New York: Oxford University Press, 2006).

[10] S. R. Weart and G. Weiss Szilard, Eds., *Leo Szilard; His Version of the Facts. Selected Recollections and Correspondence* (Cambridge, MA: MIT Press, 1978), p. 17.

"Allies"—memorial of Franklin D. Roosevelt and Winston Churchill by Lawrence Holofcener, 1995, in front of 175–177 New Bond Street, W1S

The "Allies" memorial depicting Franklin D. Roosevelt and Winston Churchill symbolize the combined efforts of the United States and the United Kingdom in WWII, which included many areas of defense, among them war-related research and development, such as the Manhattan Project and radar. In these and other projects, scientists of the two nations, among them many refugee scientists, joined forces to help defeating German Nazism, Italian Fascism, and Japanese Militarism.

Two memorial plaques of H. G. Wells, Chiltern Court, Baker Street, NW1 (left), and 13 Hanover Terrace, NW1 (right)

Back in 1914, author H. G. Wells (1866–1946) in his book, *The World Set Free*, predicted the discovery of atomic energy. Wells wrote:

> We should not only be able to use this uranium and thorium; not only should we have a source of power so potent that a man might carry in his hand the energy to light a city for a year, fight a fleet of battleships, or drive one of our giant liners across the Atlantic; but we should also have a clue that would enable us at last to quicken the process of disintegration in all the other elements, where decay is still so slow as to escape our finest measurements. Every scrap of solid matter in the world would become an available reservoir of concentrated force. Do you realize, ladies and gentlemen, what these things would mean for us? … It would mean a change in human conditions that I can only compare to the discovery of fire, that first discovery that lifted man above the brute.[11]

---

[11] H. G. Wells, *The World Set Free* (1914).

Plaque of J. Desmond Bernal, 44 Albert Street, NW1, and his photograph by and courtesy of Alan L. Mackay

J. Desmond Bernal (1901–1971) was a physicist who created a leading research center in crystallography at Birkbeck College of the University of London. However, it would be difficult to strictly compartmentalize his areas of activities. He was a visionary scientist of great authority, whose ideas inspired others to embark on pioneering studies, even resulting in Nobel Prizes. Bernal excelled at producing concepts but did not carry detailed projects to their completion. He was one of the initiators of molecular biology, especially with his employing X-ray crystallography first for the investigation of the structure of proteins. He contributed to the British war efforts and helped plan the Normandy invasion. He was interested in the social function of science and wrote comprehensive treatises about it. He was active in politics, but, sadly, his unconditional devotion to the Soviet Union clouded his judgment.

Peter Higgs in a display of King's College, Kingsway, WC2

Peter Higgs (1929–) is a theoretical physicist and Professor Emeritus at the University of Edinburgh. He did his schooling in Bristol, then at the City of London School. He graduated in physics at King's College, received an 1851 Research Fellowship, and became a PhD in 1954. He held positions at Imperial College and UCL before joining the University of Edinburgh in 1960, from which he retired in 1996. He became famous for his prediction of a fundamental particle in 1964; it is called the Higgs boson (also known as the "God particle"). Other physicists, among them François Englert and Robert Brout, made the same prediction at about the same time, independent of Higgs. The prediction was based on a fundamental theory that explained the origin of mass. After the particle was observed experimentally in 2012, Higgs and Englert received the Nobel Prize in Physics in 2013 (Brout, by then, had died, and the Nobel is awarded only to living scientists).

Chemists

"The country which is in advance of the rest of the world in Chemistry will also be foremost in wealth and in general prosperity. For the study of Chemistry is so closely bound up with our development in all kinds of industry, with the arrestment of disease, and with our success in war, that it is essential to a wealthy, healthy, and peaceful nation."[12] These are the words of Sir William Ramsay (see below) from his fin de siècle collected essays.

---

[12]William Ramsay, *Essays—Biographical and Chemical* (London: Archibald Constable & Co., 1908), p. 19.

 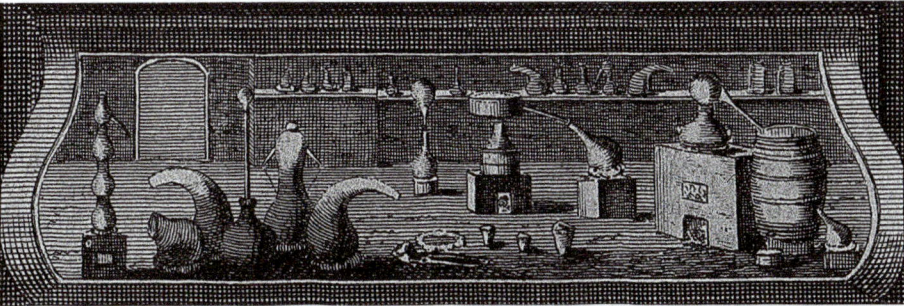

Bust of Robert Boyle in the lobby of the Royal Society of Chemistry and line engraving by B. Cole, 1753, highlighting his lab equipment (Wellcome Collection)

Robert Boyle, a pioneer of modern chemistry, was introduced earlier in this chapter.

The Royal Society of Chemistry adjacent to the Royal Academy of Art, Burlington House, and photograph of Thomas Graham by E. Edwards (Wellcome Collection)

The Royal Society of Chemistry was founded in 1841 by a group of 77 scientists and entrepreneurs. Initially founded as the Chemical Society of London, it was granted a Royal Charter 7 years later. Its first officers were Thomas Graham (1805–1869), president, and Robert Warington (1807–1867), secretary. Graham studied chemistry at the University of Glasgow and medicine at the University of Edinburgh. He was professor of chemistry in Scotland and later at UCL, having in 1837 inherited the chair of the late Edward Turner (see below). Graham invented dialysis, a process drawing upon his training in both chemistry and medicine. In 1854, he left UCL and was appointed Master of the Mint when Sir John Herschel resigned from the lucrative position once occupied by Isaac Newton.

Warington investigated the adulteration of tea; his findings assisted a parliamentary inquiry into this matter in 1855. He discovered the so-called aquarium principle, according to which plants added to water produce oxygen, thus helping support animal life, a finding still applied in present-day aquaria. Warington did some of his studies at UCL and supplemented his income by working for a brewery. Both an active research chemist and organizer within the profession, he was the driving force behind establishing the Chemical Society and its successor, the Royal College of Chemistry (more about it below).

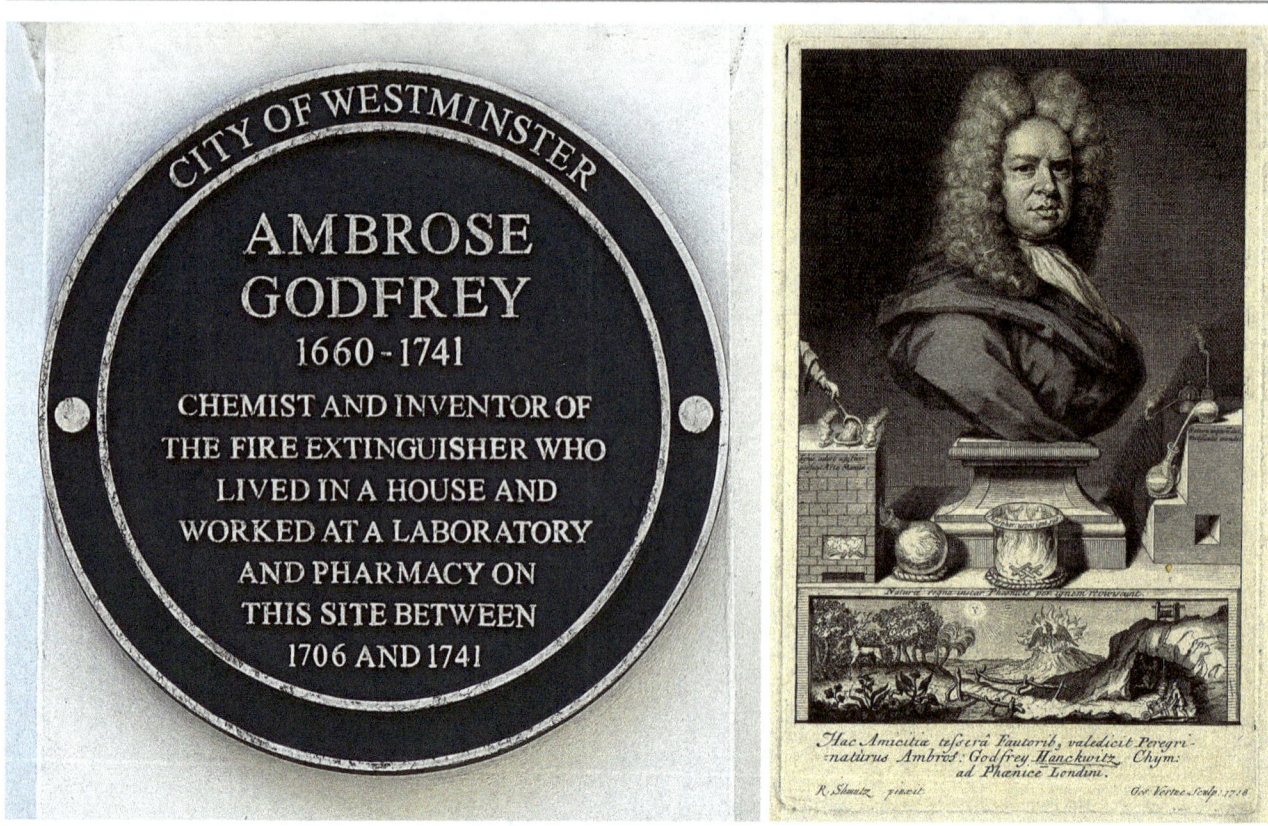

Plaque of Ambrose Godfrey, 30–32 Southampton Street, WC2, and his engraving by George Vertue, 1718, after R. Schmutz (Wellcome Collection)

Ambrose Godfrey (1660–1741) was a German-born (Gottfried Hankwitz) chemist who joined Robert Boyle's laboratory in 1679. There, another German-born chemist, Johann Becher, was trying to produce phosphorus, and Godfrey became his assistant. After Godfrey learned from a Hamburg chemist, Hennig Brandt, about the right temperature conditions, they succeeded to isolate phosphorus from urine. Having satisfied Boyle's curiosity about this element, Godfrey commercialized the process and built a booming business producing and selling phosphorus. His apothecary shop contained a separately built workshop where he experimented with other elements. He invented and patented the first fire extinguisher of record in 1727. Godfrey expressed his respect toward Boyle by naming his first-born son Boyle Godfrey.

From left to right: Imperial Chemical House, designed by Frank Baines, 9 Millbank, SW1. Today, it is the Office of Gas and Electricity Markets—a government department. Statue of Chemistry on the façade. Back of the building, Smith Square, SW1

The headquarters of Imperial Chemical Industries (ICI) was opened in 1929 after ICI was created by merging several companies, but in the 1990s, ICI left the site after restructuring. The façade has three sections: one faces Millbank; another, the western bridgehead of Lambeth Bridge; and the third, Horseferry Road. The back of the building faces Smith Square. The façades are decorated by giant niches, each dedicated to a person with a portrait carved into the keystone. Four of the chemists featured were directly associated with ICI or its predecessors, viz., Harry McGowen, Alfred Nobel, Ludwig Mond, and Alfred Mond, and they are discussed in Chap. 5. Seven others were great contributors to chemistry, and they are discussed below.

Statue of Joseph Priestley by Gilbert Bayes, 1915, above the entrance, 30 Russell Square, WC1. The building is part of University College London; originally it was the Royal Institute of Chemistry (the arrow indicates where the statue is)

 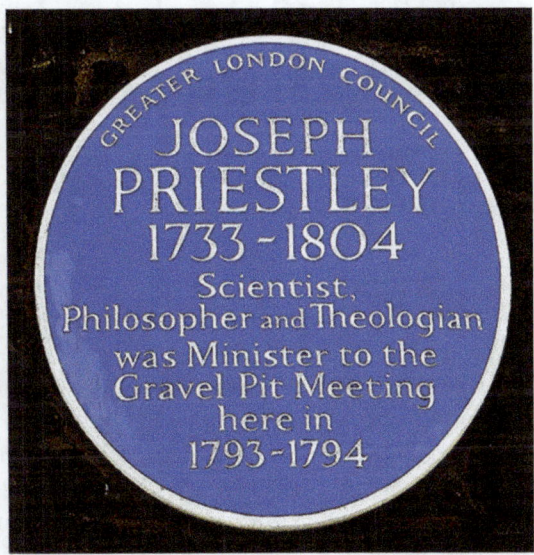

Stone portrait of Joseph Priestley on the front façade of the former ICI building and his plaque, 7–8 Ram Place, E9 (this image, Spudgun67)

Joseph Priestley (1733–1804) had an unusual life and career as a chemist, theologian-clergyman, and political thinker. His independence involved him in political controversies, and for the last 15 years of his life he had to flee his home in England and find refuge in the United States. His meeting with Benjamin Franklin moved him toward studying and writing about electricity. He invented soda water, but his greatest achievement in science was the discovery of oxygen, sometimes attributed to the French Antoine Lavoisier or the Swede Carl Wilhelm Scheele. This discovery is momentous, having been recognized as the starting point of the chemical revolution.

Priestley was a progressive thinker, but even the discovery of oxygen could not deter him from sticking to an outdated theory about the nature of air. This theory taught that combustible bodies contain a fire-like component, the phlogiston, released during combustion. Finding oxygen in air and knowing that it was combustible should have caused Priestley to question the phlogiston theory, but it did not. Priestley discovered more fundamental gases, such as nitric acid (NO), nitrous oxide (the "laughing gas," N_2O), hydrogen chloride (HCl), ammonia (NH_3), sulfur dioxide (SO_2), silicon tetrafluoride (SiF_4), nitrogen, and one that only later was identified as carbon monoxide (CO). He was a superb instrument builder as well. The blue plaque shown above refers to the few years in Priestley's life when he was Minister to the original Gravel Pit congregation, a nonconformist Presbyterian congregation named for a nearby gravel pit.

Beside the rather intriguing Priestley statue and the stone portrait in London, shown above, there are more traditional Priestley statues in Birmingham and Leeds. The highest award of the American Chemical Society is the Priestley Medal.

From left to right: Stone portrait of Henry Cavendish on the Horseferry Road façade of the former ICI building; aquatint by C. Rosenberg after W. Alexander (Wellcome Collection); and a memorial tablet, 11 Bedford Square, WC1

Henry Cavendish (1731–1810) was a natural philosopher, in today's terms, chemist and physicist. He was born in France, where his mother had gone for her health. Cavendish studied at the Newcome's School near London and at the University of Cambridge. He was reserved and shy but active in the scientific community and especially in the Royal Society. For years at the start of his career in science, he did not publish; his first paper appeared only in 1766, but many publications followed. He had collected an excellent library for his own use and a separate one for the scientific public. Keeping meticulous records, when he borrowed a book from his own library, he signed a receipt just as he would be using a public library. His experimental records and observations were both accurate and precise. From among his physics experiments, we mention the determination of the constant of gravitation. Furthermore, he investigated the properties of electricity, contributed to the theory of heat, and estimated the density of the Earth. In chemistry, he investigated atmospheric gases. He determined the composition of water and of air and deserved also the label of a pioneer of modern chemistry. His best-known achievement was the discovery of hydrogen, which he called inflammable air. When he burned it, the product was water. Antoine Lavoisier repeated the experiment and coined the name *hydrogen*. The Cavendish Laboratory at the University of Cambridge was named so by James Clerk Maxwell, the first Cavendish Professor of Physics, to honor both Henry Cavendish, the scientist, and one of Henry's relatives, William Cavendish, whose endowment supported the Laboratory.

Antoine Lavoisier on the front façade of the former ICI building and his engraving by C. F. Levachez (Wellcome Collection)

Antoine Lavoisier (1743–1794), French chemist, was one of the select few who have been called the father of modern chemistry and one of the three scientists to whom the discovery of oxygen has been attributed. He coined the names *oxygen* and *hydrogen* and contributed decisively to the development of an unambiguous chemical nomenclature. Considered to be one of the greatest scientists of all time, he recognized the conservation of mass, helped design the metric system, and described the properties of a number of elements and compounds. He transformed chemistry from a qualitative science into a quantitative one by placing emphasis on the accuracy of measurements and reporting of experimental observations. His tragedy was being embroiled with the hated and corrupt old regime at the time of the French Revolution. When the revolutionaries accused him of illegal violations of the new order, he was sentenced to death and guillotined. Not long after his execution, he was exonerated.

From left to right: Stone portrait of John Dalton on the back façade of the former ICI building; his engraving by W. H. Worthington, 1823, after J. Allen, 1814 (Wellcome Collection); and Francis Chantrey's John Dalton statue, 1838, at the Manchester Town Hall

John Dalton (1766–1844) did not receive a formal education because his parents could not afford it. Precocious in self-study, by the age of 12, he was teaching at a village school to supplement his parents' meager income. He was employed by a series of schools as a teacher or principal. In 1793 he landed a tutorship in mathematics and philosophy at the New College of Manchester. About this time, he published his first book on meteorology. Other books followed on a variety of topics. He investigated his familial condition of color blindness ("daltonism") and determined that it was hereditary. Investigating the properties of gases, he discovered the Law of Partial Pressures, according to which the total pressure of a gas mixture is the sum of the partial pressures that the component gases would exert individually if filling the same volume. Of all his works, the best known is his atomic theory and his charts of the atomic weights. He described his theory in detail in his *A New System of Chemical Philosophy* and its Appendix, published in 1810. His teachings helped the chemical industry in establishing and observing the correct proportions of the component elements in the compounds being produced.

Dalton declined the Fellowship of the Royal Society on account of his Quaker modesty, but accepted an honorary doctorate from Oxford University. He lived to see a statue by Francis Chantrey erected in his honor in 1838. It now stands in the Manchester Town Hall. Opposite this statue stands another statue of James Prescott Joule by Sir Alfred Gilbert. This closeness of the two statues is most appropriate as the elderly Dalton was a mentor of the young Joule. Joule's only memorial in London—that we are aware of—is a memorial tablet at Westminster Abbey, near the graves of Newton, Herschel, and Darwin (Joule is buried elsewhere). The tablet singles out Joule's Law of the Conservation of Energy and Joule's determination of the mechanical equivalent of heat from among his accomplishments in science.

Edward Turner (1796–1837) was a Jamaican-born British physician and chemist. His family ran plantations in Jamaica, but when he was still very small they relocated to Bath. He started schooling in Bath, attended the University of Edinburgh, and graduated as a physician. After a short stint practicing medicine, he changed professions and spent 2 years in Göttingen immersing himself in chemistry and mineralogy. When University College London (UCL) opened in 1827, he was appointed to be the first Professor of Chemistry. In concert with his interest, he was also appointed to a lectureship in geology. He did research on atomic weights and popularized John Dalton's theory of atoms. He also authored a chemistry textbook, which pioneered the usage of chemical symbols for elements and chemical formulae.

Bust of Edward Turner by Timothy Butler, 1838, in the Turner Laboratory, Chemistry Department, University College London (Wikimedia)

Justus von Liebig; from left to right: Stone portrait on the front façade of the former ICI building; his portrait by Z. Belliard (Wellcome Collection); and his bust at the Royal Society of Chemistry

Justus von Liebig (1803–1873) was a German chemist who as a teenager carried out chemical experiments in his father's laboratory. He studied at the University of Bonn and the University of Erlangen, then, under Joseph-Louis Gay-Lussac, at the University of Paris. Liebig and another German chemist, Friedrich Wöhler, discovered a fundamental phenomenon in chemistry, which later received the name isomerism by the Swedish Jöns Jacob Berzelius. Liebig defined isomers as two substances consisting of the same number and kind of atoms but with different properties. Liebig received his appointment of professorship at the University of Giessen, which was arranged for him by the great authority Alexander von Humboldt. Giessen was a small school, but it grew under Liebig's influence, and his reforms in training chemists extended all over Germany and beyond. He enhanced the independence of teaching chemistry from pharmacy, encouraged the applications of chemistry in other fields beyond medicine, and developed new technologies for the analysis of organic compounds. His laboratory attracted excellent students both in Germany and internationally. In the next generations, many leading university professors had been Liebig's students or his students' students. New laboratories were modeled after Liebig's, among them the Royal College of Chemistry in London (see more about it below). Liebig's methods of analysis facilitated the syntheses of many new organic compounds and the investigation of the degradation processes of substances of physiological significance. He demonstrated the virtually unlimited possibilities of the application of organic chemistry in agriculture and the food industry. Toward the end of his life, he moved to the University of Munich where he devoted himself to popular lecturing and writing, becoming known as the elder statesman of chemistry.

Plaque of W. Hofmann, 9 Fitzroy Square, W1, and his portrait (Wellcome Collection)

August Wilhelm von Hofmann (1818–1892) was a German chemist who studied under Justus von Liebig at the University of Giessen. He learned from von Liebig the value of an experimental approach to organic chemistry in looking for industrial applications of all new developments in the field.

Image of the first laboratories of the Royal College of Chemistry, 1846, lithograph by Day & Haghe after J. Lockyer, 1846 (Wellcome Collection) and a memorial tablet, 299 Oxford Street, W1

In 1845, von Hofmann became the first director of the Royal College of Chemistry, which began its operations on Oxford Street. Its purpose was to instruct in practical chemistry. Prince Albert supported it and laid the foundation stone for its construction. Leading politicians, including Benjamin Disraeli, William Gladstone, and Sir Robert Peel, made donations to facilitate the establishment of this institution. Sir William Crookes, J. A. R. Newlands, and Sir William Henry Perkin were among its students. The achievements of von Hofmann and his associates were part of the revolution of chemistry and chemical industry, particularly regarding the dye industry, which grew out of their discoveries in organic chemistry. The College merged into the Royal School of Mines (Chap. 5) in 1853, and in time it became the Department of Chemistry of Imperial College. When Hoffman was appointed to the University of Berlin in 1865, he helped found the German Chemical Society. British appreciation for his contribution to lifting the level of experimental chemistry and to the science of chemistry continued long after his departure and was expressed in his Copley Medal and Faraday Lectureship, both in 1875, and his Albert Medal in 1881.

research and his discoveries extended over several areas of chemistry, including some he helped pioneer, for example, metal-organic chemistry, the discovery of helium, analytical chemistry to monitor water quality in London, and the concept of valence. This latter theory was about the limitations of any atom in combining with other atoms, hence was a fundamental contribution to the fledgling structural chemistry.

Stone portrait of Marcellin Berthelot on the back façade of the former ICI building and a lithograph by J. B. A. Lafosse, 1868, after P.-L. Pierson (Wellcome Collection)

Marcellin Berthelot (1827–1907) was a French chemist, science historian, and politician. He studied in Paris and obtained a degree from the prestigious Collège de France in 1849 and his doctorate in organic chemistry in 1854. He earned a second doctorate in pharmacy in 1858. He held professorships, first at the École de Pharmacie and from 1865 at the Collège de France. He excelled in synthesizing organic compounds, and his techniques opened the way to produce broad classes of organic compounds. It was a drawback that he opposed the atomic ideas, which severely limited his interactions with the international community and his influence in the much-needed international agreements in such matters as chemical formulas and atomic weights. Over the years, his interests in chemistry broadened and he became increasingly involved with thermochemistry, the study of the relationship between heat and chemical reactions. A prolific author, he published about alchemy, philosophy, ethics, and Lavoisier. In a parallel political career, he served as minister of public education and fine arts and as minister of foreign affairs. He was active in the matters of the Science Academy and followed Louis Pasteur in the office of the Secretary of the Academy. He is buried in the Panthéon.

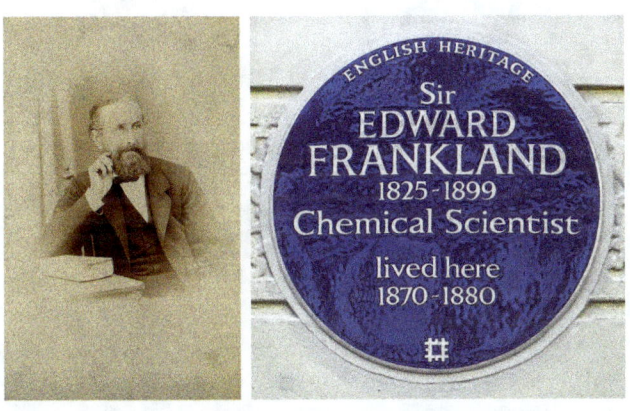

Photograph of Sir Edward Frankland by Sawyer, 1868 (Wellcome Collection), and his plaque, 14 Lancaster Gate, W2 (Spudgun67)

Sir Edward Frankland (1825–1899) was an "illegitimate" child, so he could never reveal the identity of his biological father. He apprenticed to a druggist and completed his education in chemistry in Germany. Robert Bunsen (1811–1899) in Marburg and Justus von Liebig in Giessen were among his mentors. Frankland's career included professorships of chemistry in a number of British institutions, among them, what is today the University of Manchester, St Bartholomew's Hospital, the Royal Institution, and the Royal School of Mines, where he stayed for two decades. He excelled in

Left: Dmitri I. Mendeleev's stone portrait (highlighted insert) on the central front façade of the former ICI building, facing the western bridgehead of Lambeth Bridge. Right: Plaque of J. A. R. Newlands, 19 West Square, SE11

As it often happens with significant discoveries for which the time had become ripe, several scientists at more or less the same time observed periodicity in the system of the chemical elements. The best known of them were Dmitri I. Mendeleev (1834–1907) in Russia, Lothar J. Meyer (1830–1895) in Germany, and John A. R. Newlands (1837–1898) in England. Newlands did not fare well in his home territory, and he did not expose his discovery internationally. When he revealed his observation of periodicity in the properties of the elements, some of his colleagues ridiculed him—it was so alien to them to absorb his revolutionary idea. Someone even asked him whether he had tried to classify the elements according to the initial letters of their names. Like Mendeleev, Newlands also made predictions for elements not yet known, but his ideas did not meet interest or approval. His written accounts reflect his justified bitterness. His colleagues ridiculed his discovery, but they welcomed that of Meyer and Mendeleev. Mendeleev and Meyer were awarded jointly the prestigious Davy Medal of the Royal Society in 1882 "for their discovery of the periodic relations of the atomic weights." A few years later, in 1887, Newlands also received this distinction "for his discovery of the periodic law of the chemical elements."

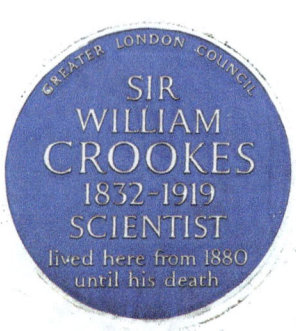 

Plaque of Sir William Crookes plaque, 7 Kensington Park Gardens, W11, and a caricature of Crookes holding a Crookes tube, by Sir Leslie Ward (Wellcome Collection)

Sir William Crookes (1832–1919) was a most colorful character in science. Although he was both a chemist and a physicist, we included him among the chemists for his discovery of the element thallium. He studied organic chemistry at the Royal College of Chemistry and was assistant to August Wilhelm von Hofmann. Crookes's meeting with Michael Faraday at the Royal Institution made him switch to optical physics. The eldest of 16 siblings, upon his father's death he inherited a substantial wealth, which allowed him to run his own physics laboratory. Still, he did not abandon chemistry, worked in chemical analysis and spectroscopy, and founded and edited the weekly *Chemical News* in 1859. It was a more informal periodical than most scientific journals and more successful, and it generated considerable income for Crookes. Between 1864 and 1879 he participated in editing the *Quarterly Journal of Science*, which also turned out to be a commercial success. He was involved in other publishing projects and among other achievements may be considered a prolific science journalist. Having invented the Crookes tubes for investigating cathode rays under vacuum, he pioneered the study of a whole family of physical phenomena, even plasma, using these vacuum tubes. He became interested in radioactivity and did analytical chemistry on uranium. In 1913, he invented a new kind of lens for eye protection that was capable of blocking 100% of ultraviolet light and 90% of infrared radiation. He was versatile and flamboyant, always interested in learning about and understanding anomalies. His positive and negative traits made him stand out among his peers. During the second half of his life, he was much taken by spiritualism and had all sorts of noisy episodes in this connection. Crookes received the highest decorations and served as President of the Royal Society. One Crookes device has never lost popularity: the Crookes radiometer, a glass bulb under some degree of vacuum with vanes mounted on a spindle within. When the bulb is exposed to light, the vanes rotate; the higher the light intensity, the faster the vanes rotate. This experiment demonstrates the conversion of the light energy via heat into motion.

Plaque of Sir Norman Lockyer, 16 Penywern Road, SW5, and his photograph by Walery (Wellcome Collection)

Sir Norman Lockyer (1836–1920) received traditional schooling. His interest as an amateur astronomer brought him a seminal discovery and an esteemed position in academia. In the 1860s, electromagnetic spectroscopy was a tool of astronomers for determining the positions of objects in the sky. In 1868 Lockyer observed a line in the spectrum of the Sun that pointed to a heretofore unknown element. He named it *helium* after the Greek word for the Sun, Helios. The French Pierre Janssen also identified this element that same year, so the two are considered co-discoverers. It took another 27 years before William Ramsay identified terrestrial helium. In 1885, Lockyer was appointed Professor of Astronomical Physics at the Royal College of Science, which was one of the predecessor components of Imperial College. Sensing the need for improved science communication, Lockyer founded *Nature* in 1869 and remained its editor until the end of his life—more than half a century.

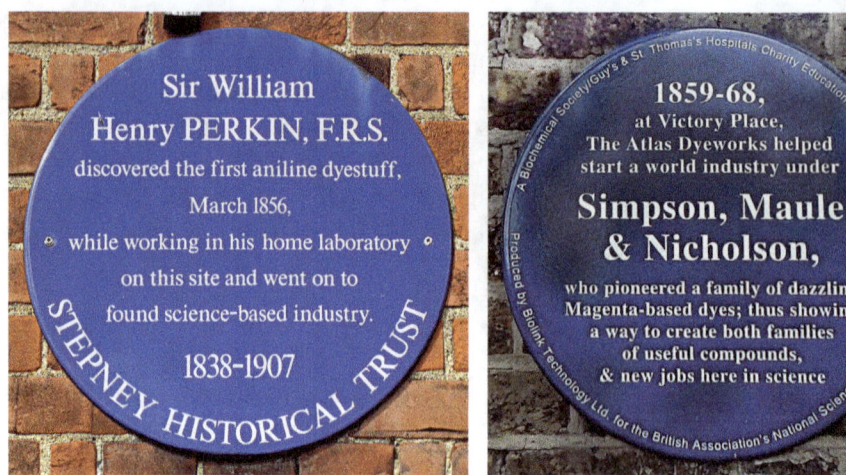

Plaque of Sir William Henry Perkin commemorating his discovery of the first aniline dyestuff in 1856, Cable Street, E1 (the corner with Sutton Street), and the plaque commemorating the dyestuff industry under Simpson, Maule, and Nicholson, 1859–1868, Victory Place (46 Rodney Road), SE17; the images are courtesy of Philip Ball

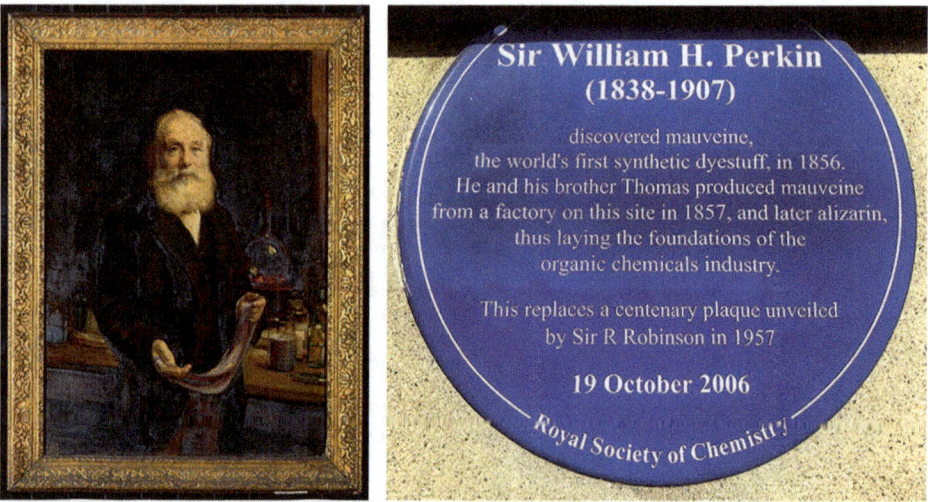

Sir William Henry Perkin in old age, painting by Arthur S. Cope, 1906 (Wellcome Collection), and plaque in Greenford Green at the site of the synthetic dyestuff factory of the Perkin brothers; courtesy of Henry Rzepa

Sir William Henry Perkin (1838–1907) could be presented among the inventors just as well. Although his main invention was independent of his mentor, Professor Hofmann, his activities emerged from their interactions. Perkin attended the City of London School when he was 14 and one year later enrolled in the Royal College of Chemistry under August Wilhelm von Hofmann, few years later becoming von Hofmann's assistant. Perkin was much taken to experimental organic chemistry, and he conducted experiments not only at the College, but also at home on Cable Street. It was there that in 1856 he discovered a purple product as a result of one of his experimentations with aniline. It was a beautiful example of how tinkering with organic substances in a kind of research setting led to a major discovery. Together with his brother and a friend, Perkin constructed a hut to serve as their laboratory, and soon realized that they discovered a way of producing a synthetic dye. They called it mauveine, filed for a patent, and embarked on its commercialization. Their product could be produced in large quantities from inexpensive starting materials and enjoyed great demand in the booming textile industry. The new dye industry became part of the industrial revolution. Both of Perkin's sons, William Henry, Jr. (1860–1929), and Arthur George (1861–1937), became chemistry professors and researchers of great renown, and both continued their father's line of interest. The memorial plaque at the site of the Perkin factory in Greenford was erected in 2006, replacing an old plaque that was erected in 1957 in celebration of the centenary of mauveine production. The 1957 monument was unveiled by the famous organic chemist and Nobel laureate Sir Robert Robinson.

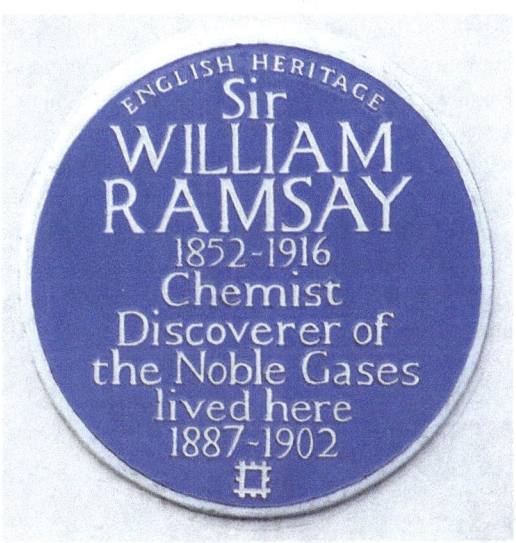  

From left to right: Plaque of Sir William Ramsay, 12 Arundel Gardens, W11; his colored photogravure by Sir L. Ward [Spy], 1908; and he and Pierre Curie in the laboratory (both, Wellcome Collection)

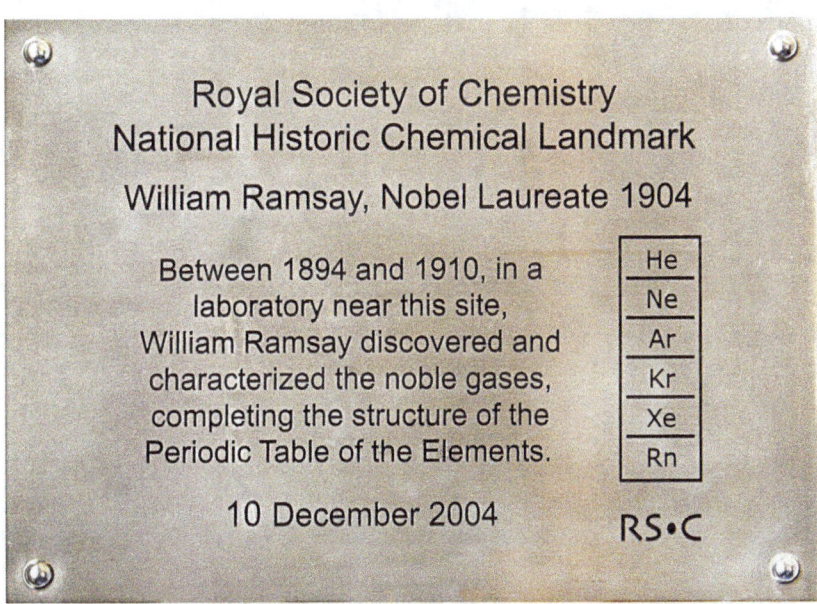 

Left: Plaque listing the noble gas elements and commemorating William Ramsay's discoveries in Slade Hall of University College London (courtesy of James Marshall). Right: Lord Rayleigh in his laboratory, photograph after Sir P. Burne-Jones (Wellcome Collection)

Sir William Ramsay (1852–1916) studied at Glasgow Academy and the University of Glasgow. He completed his chemistry training in Germany where he earned his doctorate at the University of Tübingen. Following college positions in Glasgow and Bristol, in 1887 he was appointed to the chemistry chair at UCL. In 1894, as a result of interactions with Lord Rayleigh, the two discovered argon. In subsequent years, Ramsay made a series of discoveries about what at that time were called inert gases. He and his associates discovered neon, krypton, and xenon; made the first terrestrial observation of neon; and in 1910, discovered radon. In 1904, Ramsay received the Nobel Prize in Chemistry, unshared, for the discovery of inert gases and for the determination of their place in the periodic system of the elements. The discovery of the inert gases had special significance for the Periodic Table of the Elements and, accordingly, opened the way for considering Mendeleev for a Nobel Prize. Previously Mendeleev's establishing the periodic table was considered to have been a relic of the past, but Ramsay's contribution made its prize-worthiness timely. Still, Mendeleev did not receive the Prize. Ramsay enjoyed great authority in the scientific community and beyond. Consulted about the possible location of the Indian Institute of Science, his recommendation was Bangalore.

The co-discoverer of argon, John William Strutt—Lord Rayleigh (1842–1919), was a physicist at the University of Cambridge. He also received the Nobel Prize in 1904, but for Physics, and it, like Ramsay's, was unshared. Rayleigh's investigations of gas densities were mentioned in addition to the discovery of argon. He made other fundamental discoveries, such as the theoretical interpretation of the light scattering by particles much smaller than the wavelength of light; this is known as *Rayleigh scattering*. His chief memorial is a marble tablet with a portrait relief on the wall of the St Andrew Chapel of Westminster Abbey by Francis Derwent Wood, 1921, inscribed to "John William Strutt: OM: PC: 3rd Baron Rayleigh." It mentions his chancellorship of the University of Cambridge and his presidency of the Royal Society and calls him "an unerring leader in the advancement of natural knowledge."

Plaque of Sir Frederick Gowland Hopkins, 50a The Ridgeway, Enfield, EN2 (Spudgun67), and his Photogravure (Wellcome Collection)

Sir Frederick Gowland Hopkins (1861–1947) studied at the City of London School, University of London, and Guy's Hospital. He started his research in chemical physiology, from which discipline biochemistry developed. He was the first Professor of Biochemistry at Cambridge University and one of the discoverers of vitamins, for which he received the Nobel Prize in Physiology or Medicine in 1929. The Prize was shared with the Dutch Christiaan Eijkman. The Polish Casimir Funk was conspicuously omitted although he is credited by many with the original discovery of vitamins. Hopkins popularized biochemistry when, for example, during and following WWI, the nutritional value of margarine was questioned. He determined that by adding vitamins A and D to margarine, its nutritional value could be elevated to more resemble that of butter.

Plaque of Christopher Ingold on the Department of Chemistry, UCL, 20 Gordon Street, WC1H

Sir Christopher Ingold (1893–1970) worked at the Chemistry Department of UCL between 1930 and 1970. As the text of his plaque says, he "pioneered our understanding of the electronic basis of structure, mechanism and reactivity in organic chemistry, which is fundamental to modern-day chemistry." He was a rather unusual mentor. He had little interaction with his disciples, but was aware of the progress of their research and paved their way for advancement. One of his former associates of world renown, Ronald J. Gillespie (1924–), told us that Ingold wrote up his associates' results in a series of papers and published them under the name of the young colleague alone.[13] This selflessness was quite different from the attitude of many other mentors who added their own names to the papers written up by their associates. Ingold chose students for his younger associates and made arrangements for many students to receive grants and instrumentation. At the same time, beyond a certain degree, such care may become an impediment to independence.

---

[13] I. Hargittai, *Candid Science III: More Conversations with Famous Chemists* (edited by Magdolna Hargittai, London: Imperial College Press, 2003), Chapter 4, "Ronald J. Gillespie," pp. 48–57.

Group portrait (Wellcome Collection): Back row, left to right: Selman A. Waksman, Howard Florey, Jacques Tréfouël, Ernst B. Chain, André Gratia; front row, left to right: Pierre Fredericq and Maurice Welsch; and plaque of Ernst B. Chain, 9 North View, SW19 (Spudgun67)

Sir Ernst B. Chain (1906–1979) was a German-Jewish refugee who left Germany in time to avoid persecution, but his mother and sister were murdered by the Nazis. Having received his first degree in chemistry in Berlin, Chain earned his PhD degree in Cambridge under Sir Frederick Gowland Hopkins's mentorship. From 1939, he worked on penicillin and devised the technology for obtaining pure penicillin from large quantities of mold broth. He was a co-recipient of the Nobel Prize in Physiology or Medicine in 1945 (more about this in Chap. 4). Chain moved to Rome after WWII and continued his research of penicillin production. He returned to London in 1964, joined Imperial College, and founded the Department of Biochemistry where he held its chair until his retirement in 1976. Under Chain's tenure, Imperial College became a world center in biochemistry and the Department received a seven-story new building. It is now home of the Department of Life Sciences and is named Sir Ernst Chain Building-Wolfson Laboratories. The dedication in 2012 included the unveiling of a bust of Chain, created and donated to the College by the sculptor Oscar Nemon. Back in the early 1940s, Nemon's wife had been one of the first patients whose life was saved by a sample of Chain's penicillin. When the desperate Nemon turned to Chain for help, the antibiotic was not yet released to be administered to patients. This is what Nemon remembered Chain told him: "Penicillin's my child, and I'm not allowed to see it, touch it. But what I'll do is this—I'll steal it!"[14]

---

[14] Natasha Martineau, "Sir Ernst Chain is honoured in building naming ceremony" (05 November 2012), https://www.imperial.ac.uk/news/115787/sir-ernst-chain-honoured-building-naming/

Bust of Elsie Widdowson by Margo Bulman (1974) at Imperial College, Exhibition Road, SW7 (the original is at the Royal Society)

Elsie Widdowson (1906–2000) was one of the first woman graduates of Imperial College. She had a brilliant carrier in science in a number of diverse areas, such as veterinary medicine, infant physiology, farm practices, and, above all, in nutrition. She analyzed the chemical composition of food and made recommendations for improving diet during WWII and also in peace time, in particular, fighting malnourishment in Africa. She and fellow nutritionist Robert McCance (1898–1993) formed a most successful, 60-year-long research partnership.

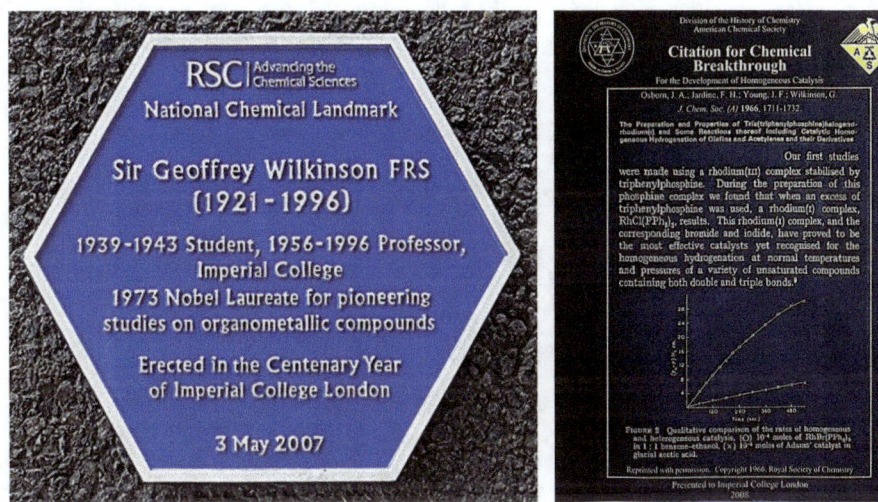

Plaque of Sir Geoffrey Wilkinson on the façade of the Sir Ernst Chain Building and a tablet honoring one of his seminal papers [*Journal of the Chemical Society* (A), 1966, 1711–1732]

Sir Geoffrey Wilkinson (1912–1996) worked in defense projects of allied British-Canadian units between his university degrees from Imperial College. His professional career included stints at Berkeley and at Harvard University before returning to Imperial College. His research interest was in catalysts; in particular, he was a pioneer in metal-organic compounds, whose best-known representative is ferrocene, $(C_5H_5)_2Fe$. The German Ernst Otto Fischer worked independently in the same area. The two shared the 1973 chemistry Nobel Prize for their discoveries.

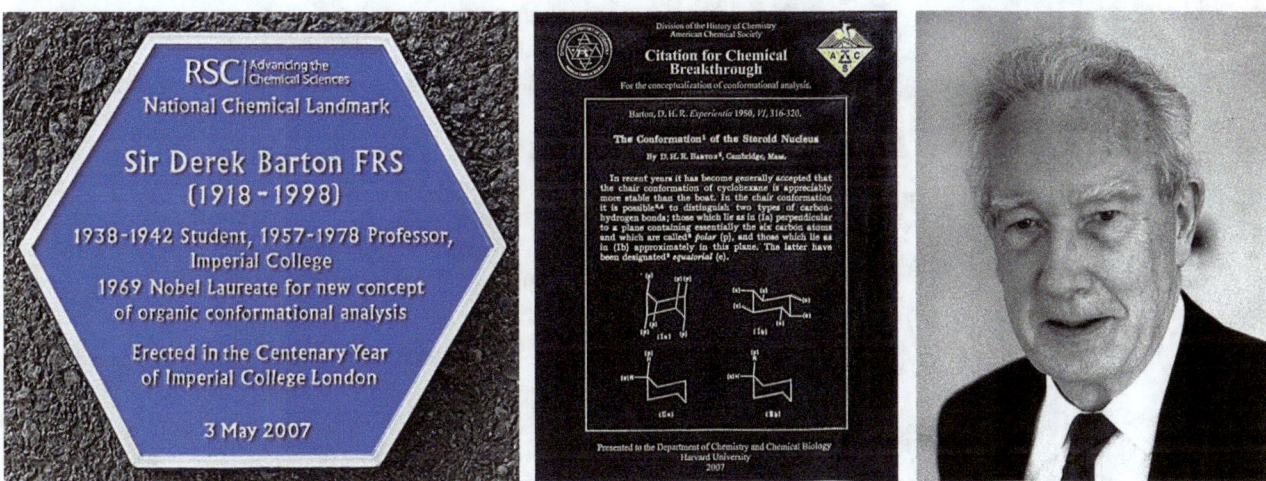

From left to right: Plaque of Sir Derek H. R. Barton on the façade of the Sir Ernst Chain Building; a tablet honoring Barton's seminal paper [*Experientia* 1950, VI, 316–320]; and his photograph at a meeting in his honor in 1988 in Austin, Texas

Sir Derek Barton (1918–1998) received his university degrees in chemistry at Imperial College, where he spent most of his professional career in the British period of his life. Then, there was a French period, 1978–1986, at the Institut de Chimie des Substances Naturelles in Gif-sur-Yvette. Finally, he spent his last dozen years at Texas A&M University. He was an organic chemist who discovered new reactions, some of which have been named after him. In 1950 he showed the importance of rotational forms, called conformers, in many substances occurring in nature. The fundamental concept of conformers was discovered by the Norwegian chemist Odd Hassel (1897–1981), but Barton demonstrated its practical implications. The two shared the 1969 chemistry Nobel Prize.

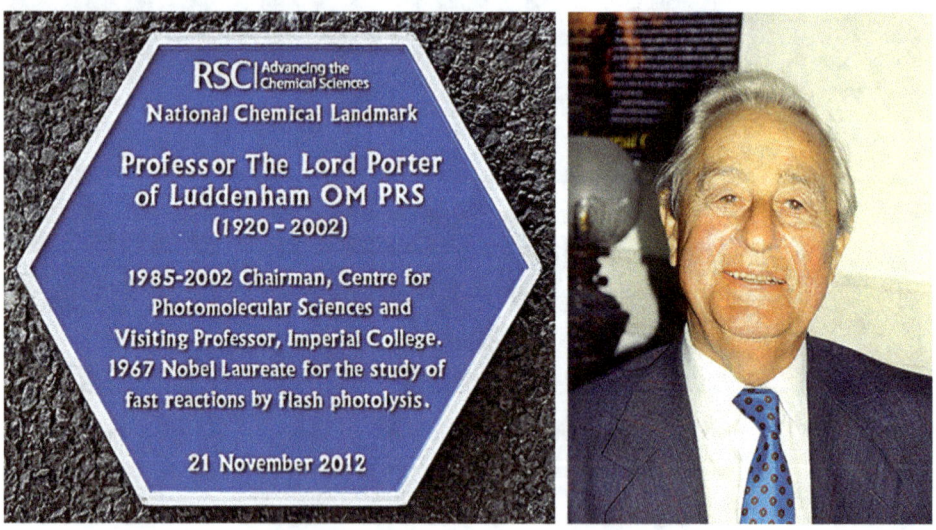

Memorial plaque of Lord Porter on the façade of the Sir Ernst Chain Building and his photograph in 1997 in his office at Imperial College

For more on George Porter, see earlier in this chapter.

Plaque of Dorothy L. Sayers, 23–24 Great James Street, WC1

Lord Kelvin figures in this book among the physicists (above) and also among the inventors (Chap. 5). Still, it is appropriate to remember him among the chemists for his pivotal contribution to understanding the concept of chirality. Lord Kelvin applied for the first time the terms *chiral* and *chirality*. Our hands, for example, are chiral; they have chirality. The example is more than apt as the term *chirality* is derived from the Greek word for hand. Our left hand and right hand, in an ideal case, have exactly the same measures, yet the two are not superimposable; this is chirality. Likewise, the molecules of many substances may occur in the two versions, left-handed and right-handed. However, there is a fundamental difference between the molecules found in nature and those produced in the laboratory. In nature, the molecules occur as a mixture of the left-handed and right-handed versions. In the laboratory, one of the two versions is produced.

In Dorothy L. Sayers's (1893–1957) detective story, *The Documents in the Case*, an expert identifier of wild mushrooms dies of poisoning. The question is, was this death a consequence of accidental poisoning or even suicide—or was it intentional poisoning, that is, murder? The analysis of the toxin in the victim's body shows the presence of only one of the two possible versions of the poisonous molecule, pointing to a man-made product, ergo, murder. Sayers had a co-author for this book, Robert Eustace (real name Eustace Robert Barton, 1854–1943). He was a physician and himself a prolific author whose plots often included references to scientific innovation. Sayers's book has been kept in print ever since it appeared in 1930. The date of the original publication is significant as it anticipated by decades the current legislation that rigorously prescribes chiral purity for pharmaceutical products. Pharmacology necessitated this legislation because the physiological impact of left-handed and right-handed versions of the same molecule may vastly differ. This is not what Sayers's story exposed, but what it exposed, some physiological relevance of chirality, was sufficiently pioneering to deserve our appreciation.

DNA—Multidisciplinary Science

The discovery of the double helix structure of deoxyribonucleic acid (DNA) has been lauded as one of the greatest discoveries in biology. It was also a discovery in structural chemistry, which utilized X-ray diffraction, model building, and the accumulated information on chemical structures.

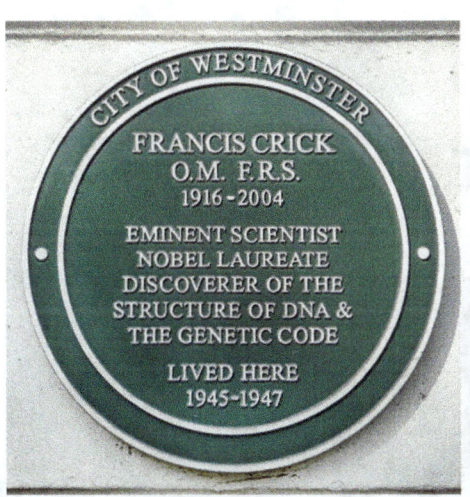

Memorial plaques of Francis Crick, 56 St George's Square, SW1 (left), and James D. Watson, 18 Vincent Square, City of Westminster (right)

Francis Crick (1916–2004) and James D. Watson (1928–) were the co-discoverers of the double helix structure of DNA, which brought about a most spectacular paradigm change in modern biology and biomedicine. The first report that DNA is the substance of heredity appeared in 1944, but this became generally accepted only thanks to further experimental studies in the early 1950s. The suggestion of the double helix structure of DNA in 1953 was key to understanding the relationship between geometrical structure and biological function. This was followed eventually by the discovery of the genetic code and by the Human Genome Project and its implications for human medicine.

Crick's plaque refers to the location where Crick lived after the war in the period 1945–1947, while he was still in the employment of Admiralty. At this time, he was looking for a worthy research project, and here he received useful advice from the Nobel laureate physiologist A. V. Hill (Chap. 1). Watson came for visits to London during the years 1983–1992, as indicated by his plaque.

Tablet recognizes the DNA scientists of King's College

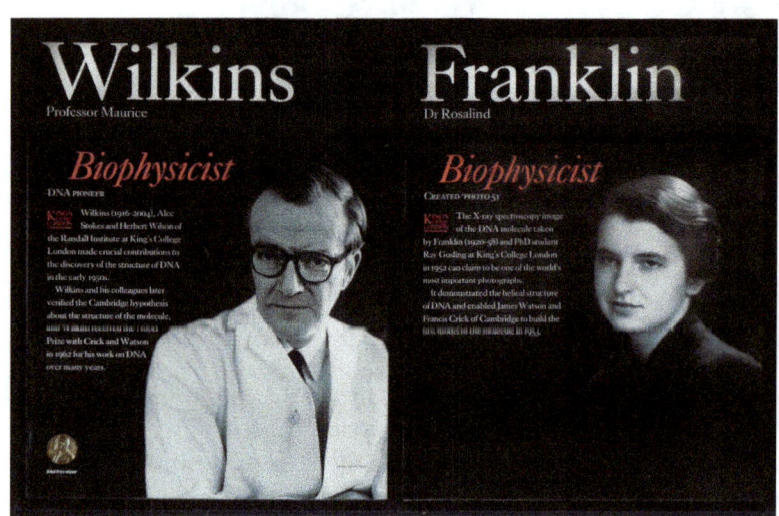

Left: Photographs of Maurice Wilkins and Rosalind Franklin displayed at the Strand Campus of King's College. Right: Plaque at King's College of R. E. Franklin, R. G. Gosling, A. R. Stokes, M. H. F. Wilkins, and H. R. Wilson

The above images honor the scientists who worked on the structure of DNA at King's College, using X-ray crystallography. Rosalind E. Franklin (1920–1958) was solving the structure in a stepwise manner, but her involvement with DNA was cut short by some controversies between her and Maurice Wilkins. She was an expert on X-ray crystallography, and she demonstrated her acumen before and after her relatively brief involvement with the DNA project. Franklin and her student, Raymond Gosling, prepared the most beautiful and most informative X-ray photograph of DNA, which might have led them to the correct structure. Wilkins showed this photograph to Watson, without Franklin's knowledge, and this insight greatly facilitated Watson and Crick's model-building and their arrival at the right solution. After Franklin had left King's College, Wilkins continued his X-ray diffraction studies of DNA and produced a large body of details about its structure. By the time the discovery of the DNA structure was deemed Nobel Prize-worthy, Franklin was no longer alive, so the award was shared among Crick, Watson, and Wilkins in 1962.

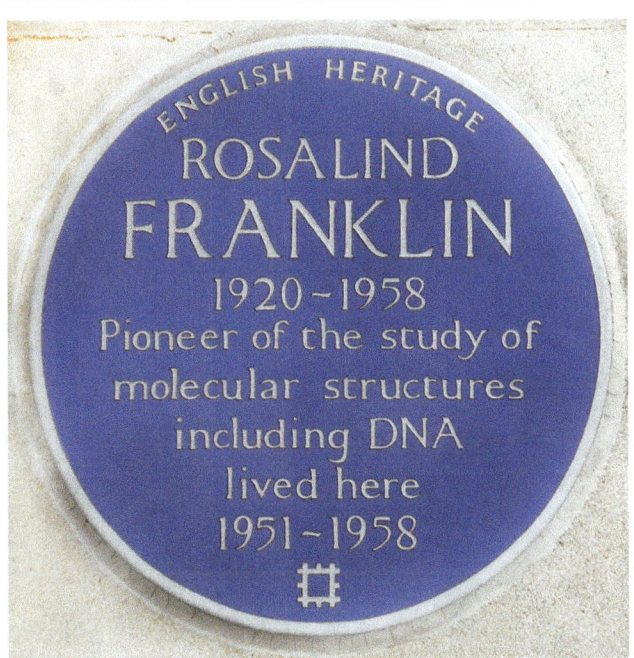

Plaque of Rosalind Franklin, Donovan Court, 107 Drayton Gardens, SW10

For some time, Rosalind E. Franklin's contribution to the discovery of the double helix structure was underestimated until her negative portrayal in Watson's book *The Double Helix* turned attention to the question of justice. Then, her former student, Aaron Klug, having examined Franklin's lab diary, concluded that she was much closer to solving the structure than had been supposed. Her findings would have come out probably in steps, which was the traditional way of publishing the results of such complex experimental investigation. Still her excellent X-ray photographs and observations communicated to Watson and Crick, without her knowledge, played a crucial role in the discovery of the double helix structure of DNA.

Raymond G. Gosling (1926–2015) worked on his PhD on the X-ray diffraction of DNA. Together with Rosalind Franklin, he produced the best X-ray diffraction pattern, "Photograph 51" of the wet B-form of crystalline DNA. It proved to be crucial in Watson and Crick's arriving at their conclusion of the double helix structure of DNA.

Alexander R. Stokes (1919–2003) was a PhD crystallographer who earned his degree at Trinity College in Cambridge under W. Lawrence Bragg. Stokes predicted what the X-ray diffraction pattern of a helical structure should be like, on the basis of mathematical considerations.

Maurice H. F. Wilkins (1916–2004) was a physicist and molecular biologist who used X-ray crystallography to determine the structure of DNA. Following the initial results and Watson and Crick's prediction of a double helix structure of DNA, Wilkins continued his painstaking X-ray diffraction work until reliable evidence was produced validating Watson and Crick's suggestion for the structure. He was one of the three, along with Watson and Crick, who shared the Nobel Prize in Physiology or Medicine in 1962 for the discovery of the DNA structure.

Herbert R. Wilson (1929–2008) worked with Wilkins on the X-ray diffraction elucidation of the structure of the A-form of crystalline DNA.

Crick and Watson eminently complemented each other in Cambridge during their quest for the DNA structure in the early 1950s. However, they were conspicuously different. Crick was a scientists' scientist; Watson was more of a public persona. Both were curious and ambitious, but Crick was more curious than ambitious, and Watson was more ambitious than curious. Crick found great challenges in science after DNA. Watson made sure that he was identified with DNA and not only with its structure. He created his own institution from the once-dilapidated Cold Spring Harbor Laboratory in the United States, excelled in science administration, and became a celebrated author.

The Francis Crick Institute at 1 Midland Road, NW1, and the "Paradigm" sculpture by Conrad Shawcross, 2016

The largest biomedical research center in the United Kingdom was named after Crick. The Crick Institute was established in 2010 from a partnership between six of the most significant institutions on world scale, viz., Cancer Research UK, Imperial College London, King's College London, the Medical Research Council, UCL, and the Wellcome Trust, and it opened in 2016. It stands next to the British Library (Chap. 1); that is, they have adjacent sites, although their street addresses are different. The sculpture in front of the Crick Institute is called "Paradigm." It consists of tetrahedra that twist as being stacked onto each other, reaching the height of 14 meters. It relates to the theory of Thomas Kuhn, according to which scientific progress happens in jumps rather than in a gradual fashion. The growing size of the tetrahedra as the structure is ascending conveys a feeling of uncertainty and skepticism.

The Science Museum displays Crick and Watson's reconstructed model of the double helix, which contains some of the components of the original model. The cafeteria in the Museum has a cartoon on the wall depicting the two discoverers with their model.

Naturalists and Biologists

Bust of John Ray by Louis-François Roubiliac at the British Museum

Statue of Carl Linnaeus by Patrick McDovell in a niche at the ground level of the east wing, among "illustrious foreigners," Burlington House

John Ray (1627–1705) was a pioneer naturalist whose main interest was in taxonomy. He introduced species as the unit of taxonomy comprising organisms that share common characteristics and are capable of interbreeding. Ray studied anatomy and chemistry at the University of Cambridge and stayed on for many years. He lost his fellowship in 1662 when he refused to take the loyalty oath required at the time of Restoration. For the rest of Ray's life, well-to-do friends supported him so that he could continue as a naturalist. He made an agreement with a fellow naturalist, Francis Willughby (1635–1672), that he would study the flora and Willughby the fauna. With great ambition, they were aiming at covering the entire natural history of all plant and animal kingdoms. Ray carried out his investigations well beyond Britain, to the European continent. The Royal Society financed his publications resulting from his studies.

Carl (Carolus) Linnaeus (Carl von Linné, 1707–1778) was a Swedish naturalist and biologist who created the modern system of classification and nomenclature of organisms. He studied at Uppsala University, where he later became Professor of Medicine and was in charge of its botanical garden. He classified animals, plants, and even minerals. His fundamental *Systema Naturae*, appearing in 1735, established his authority among European naturalists. One of his side interests was manifested in his advocacy of breast-feeding of babies by their mothers rather than employing so-called wet nurses for feeding them. In 1753, Linnaeus started publishing his *Species Plantarum*, the system of botanical nomenclature. His memorials abound in Sweden and internationally.

Left: Statue of Sir Joseph Banks at the Natural History Museum. Right: Stipple engraving by Ridley, 1802, after J. Russell, 1788 (Wellcome Collection)

Busts of Sir Joseph Banks by Anne Seymour Damer (left) and by Sir Francis Chantrey (right) in the British Museum

Sir Joseph Banks (1743–1820) was a naturalist, explorer, collector of plants, and a politician of science. He has memorials in at least three major venues connected with science and knowledge in London, viz., the British Museum, the Natural History Museum, and the British Library. He studied natural history at the University of Oxford. Soon after, he participated in his first expedition to Newfoundland and Labrador when he was only 23 years old. Most famously, he participated in Captain Cook's circumnavigating the globe with *Endeavour*, 1768–1771. Not only did Banks pay his own way, but he financed the participation of eight others, made observations, and also collected plants during this voyage. Upon his return to London, he quickly became a leading personality in British scientific life. He had already been elected Fellow of the Royal Society in 1766, and he became president in 1778, staying in this position till the end of his life. Nobody before him or since served for such a long period as president (currently the presidents are elected for a single 5-year term). He was rather autocratic in his presidency, but he elevated and strengthened the position of the Royal Society in British society as well as its authority internationally. During his tenure, the Society became increasingly the advisor of the government and he personally advisor to King George III.

Left: Stone plaque of Sir Joseph Banks, Robert Brown, David Don, and the Linnean Society, 32 Soho Square, W1. Right: Lithograph of Robert Brown by T. H. Maguire, 1850 (Wellcome Collection)

The memorial plaque at Soho Square commemorates not only the botanists Sir Joseph Banks, Robert Brown (1773–1857), and David Don (1799–1841) but also the Linnean Society, which was established in 1788 and is known as the world's oldest biological society. Charles Darwin and Alfred Russel Wallace were its most famous members. Their theory of evolution by natural selection was presented for the first time at a meeting of the Linnean Society in 1858. The current location of the Society is at Burlington House. David Don was Professor of Botany at King's College and afterward became the librarian at the Linnean Society.

Robert Brown was a prominent member of the Linnean Society. As a botanist, he participated in the expedition of Matthew Flinders to Australia. Brown's discovery of the Brown movement (or motion), named after him, is well known. This phenomenon is the random motion of particles suspended in a liquid or a gas. In 1905, Albert Einstein (and, independently, the Polish physicist Marian Smoluchowski) produced a quantitative theory of the Brownian movement. It was Einstein's first seminal contribution to physics. His interpretation of the movement of particles (pollen particles in Brown's observations under the microscope) as being moved by water molecules was considered to be evidence that atoms and molecules exist. Soon, in 1908, the French physicist Jean Perrin verified Einstein's theory experimentally. This was among Perrin's observations demonstrating the discontinuous structure of matter for which he was awarded the 1926 Nobel Prize in Physics. Thus Brown's discovery reverberated in fundamental studies at the dawn of modern physics.

Statue of Georges Cuvier by Patrick McDovell in a niche at the ground level of the east wing, among "illustrious foreigners," Burlington House, and portrait of Georges Cuvier, holding a fish fossil (Wellcome Collection)

Georges Cuvier (1769–1832) was a French naturalist and zoologist, a pioneer of comparative anatomy and paleontology, who made revolutionary discoveries by comparing living animals and fossils. He was well read from early childhood, and already at 10 years of age, he decided to dedicate himself to natural history. He studied in Germany and France and from 1795 lived in Paris. He interacted with the leading naturalists of his time and was elected to be a member of the fledgling Academy of Sciences of the Institut de France. He became professor of natural history at the prestigious Collège de France and received other distinctions, among them a foreign membership of the Royal Society. His research activities and discoveries extended to geology and other areas beside paleontology and zoology. He examined fossils on different continents and determined that some of them belonged to organisms he named as *Mastodon*, *Megatherium*, *Pterodactylus*, and *Mosasaurus*. He was the first who suggested that in prehistoric times the earth was populated mostly by reptiles and not mammals as had been supposed before. He opposed the pre-Darwinian theories of evolution that were around in his time.

Lithographs of Sir William Jackson Hooker (left) and Sir Joseph Dalton Hooker (right), both by T. H. Maguire, 1851 (Wellcome Collection), and their joint plaque, 49 Kew Green, TW9 (Spudgun67)

Sir William Jackson Hooker (1785–1865) was a botanist, Regius Professor of Botany at Glasgow University, and Director of the Royal Botanic Gardens, Kew, TW9. Early on he became interested in ornithology and entomology and generally in natural history. Eventually, he narrowed his interest to botany. He was a friend of Sir Joseph Banks, at whose suggestion he embarked on expeditions that took him to Iceland and other voyages to continental Europe with the purpose of extending his botanical knowledge and collection. He developed an internationally renowned herbarium. In the directorship of the Royal Botanical Garden, Kew, he was succeeded by his son, Sir Joseph Dalton Hooker (1817–1911) who excelled in this function at least as much as his father. Sir Joseph studied medicine at the University of Glasgow and attended his father's lectures. He, too, became a dedicated naturalist. He participated in Captain James Clark Ross's Antarctic expedition, 1839–1843. He made further voyages of exploration to India and the Himalayas, Palestine, Morocco, and to the Western United States. He was a friend and supporter of Charles Darwin and helped in classifying the plants that Darwin had collected during his travels. He was also a good friend of Thomas Huxley, another enthusiastic defender of Darwin's theory of evolution. For his original studies and discoveries about the geographical distribution of plants, Joseph Hooker is considered to be the founder of geographical botany.

Grant Museum of Zoology and Comparative Anatomy, 21 University Street, WC1E, and lithograph of Robert Edmond Grant by T. H. Maguire, 1852 (Wellcome Collection)

Robert Edmond Grant (1793–1874) initiated in 1827 what is today the Grant Museum of Zoology and Comparative Anatomy. The purpose was assisting the teaching at the then-new University of London, now University College London (UCL). The Museum can be considered Grant's memorial. He studied medicine at the University of Edinburgh and obtained his MD in 1814. He did not practice medicine; rather, he engaged in research in marine biology and the zoology of invertebrates and taught at the University of Edinburgh. The young Charles Darwin was one of his students and attended his lectures with increasing enthusiasm. Darwin became quite attached to Grant, helped him in collecting and preparing his specimens, showed keenness for doing research, and started his scientific activities as Grant's disciple. Eventually, Grant moved to London, and in 1827 he was appointed Professor of Comparative Anatomy. He remained in this position until the end of his life. He had other prestigious functions at the Royal Institution and the British Museum. Among his multifaceted activities, one was his support of Thomas Wakley's fledgling medical journal *The Lancet* (Chap. 4). At that time, it was far from the prestigious and well-established journal it has become. Wakley appreciated Grant's backing and reciprocated by publishing Grant's 60-lecture course in comparative anatomy in the *Lancet* during 1833–1834. The series was quite an undertaking but was well received.

Mary Anning (1799–1847) became an enthusiastic and knowledgeable amateur paleontologist. Among her significant finds is a complete ichthyosaur skeleton in the Blue Lias rocks of Charmouth beach in southwest England. The fossils she discovered have become the subject of serious research by later scientists. She did a great service for science, and we can only imagine what she might have become had not the situation of women in science been so restrictive in her time.

Portrait of Mary Anning by an unknown artist at the Natural History Museum

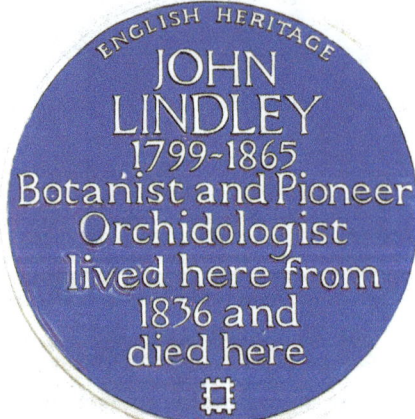  

Left: Plaque of John Lindley at Bedford House, The Avenue, W4 (Spudgun67), and his lithograph by T. H. Maguire, 1849 (Wellcome Collection). Right: Plaque of George Bentham, 25 Wilton Place, SW1X

John Lindley (1799–1865) could not afford higher education, so he found employment in the seed trade. This was somewhat connected with his father's occupation. His father ran a nursery garden, but was not very successful in business. But John Lindley did not give up his dream to study and, by luck, met Sir William Jackson Hooker, who allowed him to use the rich family library that specialized in botany. When Hooker introduced Lindley to Sir Joseph Banks, Banks recognized Lindley's thirst for knowledge and his dedication to botany. Consequently, Banks employed Lindley as assistant in his herbarium. Lindley excelled in his new circumstances, embarked in research, and started publishing. Banks died soon, but by then Lindley's career had taken off. The Linnean Society supported his efforts and elected him a Fellow and so did other learned societies. Eventually Lindley was appointed to a professorship of botany at UCL and lectured at the Royal Institution. Beside recognition in Britain, he received awards and honors internationally. Considering his humble beginnings and the hurdles he had to overcome, he had a spectacular career. This was yet another example that science may provide a better chance for talent to cut through barriers in society than in other areas of human endeavor.

George Bentham (1800–1884) became interested in botany thanks to a book he read as a child. He graduated in law, but moved to botany. Because of his interest in plants, he became interested in broader questions of natural philosophy. Through his close friend, Sir Joseph Dalton Hooker, he was introduced to Darwin's theory of evolution and to Darwin himself. Bentham developed into a strong supporter of the theory of evolution, and his accomplishments in botany and organizational skills earned him recognitions at home and internationally.

Bust of Sir Richard Owen, 1860, and his statue by T. Brock, 1897, both at the Natural History Museum, and his photograph by Maull & Polyblank (Wellcome Collection)

Sir Richard Owen (1804–1892) was a biologist who excelled in comparative anatomy and paleontology and in interpreting fossils. He coined the word *Dinosauria*. Although he agreed that evolution did take place, he was a fierce critic of Darwin's theory that it occurred through natural selection. Owen's views, though, may appear more progressive today than before in light of the recent findings of evolutionary developmental biology. Owen studied medicine, first at the University of Edinburgh, and completed his studies at St Bartholomew's Hospital. During his professional career he was connected with the Hunterian Collection, the Royal College of Surgeons, and the British Museum. There, he was the superintendent of the natural history collections. His longest lasting contribution was his advocacy for a new home for the natural specimens of the British Museum. The result was the magnificent Natural History Museum. According to contemporary reports, Owen had a rather unpleasant personality and his peers characterized him with the most awful adjectives.

Portrait of Charles Darwin by T. H. Maguire and title page of *The Origin of Species*, 1859 (both, Wellcome Collection)

Charles Darwin: bust at the Darwin Building of University College London and a plaque, both on Gower Street, WC1E

Charles Darwin (1809–1882) is one of the best-known scientists in the world for his theory of evolution by natural selection. He studied medicine at the University of Edinburgh, which was an excellent venue of science education. He then transferred to Cambridge and continued there in theology, but did not complete his studies before he enrolled for the famous expedition of the *Beagle*. He paid for his voyage; in other words, he was not an employee but a companion of the captain, Robert Fitzroy. The circumnavigation of the globe, 1831–1836, resulted in Darwin's most valuable diary and the catalogs of his enormous collection of specimens. The publication of his diary in 1839 brought him exposure and fame. In the meantime, he started working on his theory of evolution. He prepared his first draft in 1842 and expanded it in 1844 with no intention for immediate publication. He understood it was too radical and too revolutionary for the moment. Yet he kept working on what would become his magnum opus, *On the Origin of Species by Means of Natural Selection or the Preservation of Favoured Races in the Struggle for Life*. Originally, it was to be an abstract of his three-volume treatise *Natural Selection*. The abstract grew into a book and turned out to be more accessible than the longer work it was supposed to abstract. During these years, Darwin had close friends to consult with, such as the geologist Charles Lyell and the biologists Thomas Huxley and Joseph Hooker. It was not until 1859 that he and his friends read joint excerpts from his work and the work of Alfred Russel Wallace (see below) at the Linnean Society. It turned out that Wallace had come to similar ideas as Darwin and had sent his studies to Darwin.

The involvement of Alfred Russel Wallace (1823–1913) made Darwin's coming out mandatory if he did not want to lose any claim to priority. Note that the word *evolution* did not appear in the title of Darwin's 1859 publication. That term was far too challenging and was added to the title only in 1872. Darwin was sensitive to the expected public reaction to his revolutionary ideas. He loathed public debate in which Thomas Huxley stood in as a willing proxy for him. When the *Origin of Species* came out, it was far from generally accepted, but there were some even beyond the community of scientists who recognized its significance. In 1859, the incoming Liberal Party Prime Minister Lord Palmerston proposed that Queen Victoria confer Darwin a knighthood. Even though Prince Albert supported the proposal, the idea was not carried through when Bishop Wilberforce opposed it on religious grounds. Thus Darwin's highest recognition did not come from State or Crown; it came from the Royal Society, when in 1864, he received its highest award, the Copley Medal. Darwin's scientific production extended much beyond his theory of evolution. It was a milestone in progress in biology, in all of life sciences, for many decades to come. Only the discovery of DNA as the carrier of heredity and its double-helical structure would be a radical quantum leap of comparable magnitude in the middle of the twentieth century.

From left to right: Photograph of Alfred Russel Wallace by Sims, 1889 (Wellcome Collection); his statue by Anthony Smith, 1908, at the Natural History Museum; and a plaque, 44 St Peter's Road, South Croydon, CR0 (Spudgun67)

Wallace had to terminate his schooling at the age of 14 due to the financial hardship of his family. He apprenticed for 6 years, but used every opportunity to read books and attend lectures to quench his thirst for knowledge. His self-education landed him a private job of teaching drawing, mapmaking, and surveying. He continued learning and soon started publishing scholarly papers about his observations in zoology. He ventured into civil engineering and even into designing buildings. He read deeply in geology and natural history, which greatly impacted his interest and imagination. After reading Darwin's *The Voyage of the Beagle*, Wallace became a self-made naturalist. He used his savings to travel to Brazil where he investigated the Amazon River basin. Then, he traveled to the Malayan Archipelago and continued his work there. He was the founder of what would in the future be called *biogeography*, the study of the geographical distribution of animal species. He covered a broad spectrum of topics, among them even the question whether life was possible on the planet Mars. His most remarkable work was the investigation of the changes of species. Following 10 years of study, by 1858 he devised his theory of natural selection and evolution. He composed an essay and sent it to Darwin. He did not know about Darwin's similar work and conclusion, but he guessed, and he rightly surmised, that

Darwin would be the person for him to approach with his ideas. We have seen above that Wallace's study stimulated Darwin in his coming out with his own theory. As an example for posterity of fairness in publication, Darwin and his friends arranged for joint readings of excerpts from Darwin's and Wallace's writings at the Linnean Society. The two men became lifelong friends. In contrast to the wealth of Darwin and his friends, Wallace had no financial background that would allow him a worry-free engagement with science. However, from 1881, he received a modest government pension thanks to Darwin's efforts.

Left: Memorial of Henry Fawcett on the Victoria Embankment Gardens, WC2. Mary Grant was the sculptor and Basil Champneys the architect, 1886. Right: Photograph by Lock & Whitfield (Wellcome Collection)

Henry Fawcett (1833–1884) was an economist trained at King's College School in land surveying and at the University of Cambridge. A shooting accident blinded him at the age of 25. His claim to fame was his fierce defense of Darwin's theory of evolution. Fawcett participated in famous debates about evolution in Oxford in 1860 and in Manchester in 1861 at a meeting of the British Science Association. His rational arguments were a strong influence in the acceptance of Darwin's teachings.

 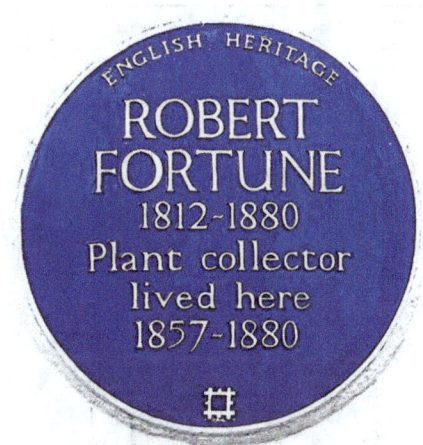

Plaques of Philip Henry Gosse, 56 Mortimer Road, De Beauvoir Town, N1, and Robert Fortune, 9 Gilston Road, SW10 (both, Spudgun67)

Philip Henry Gosse (1810–1888) was a naturalist, an early marine biologist, and an early conservationist. He was taught at home, but had to start working early. He traveled from England to Canada and became a self-taught entomologist. He was engaged in farming, taught in school, and had employment as a private tutor in Canada and in the United States. His strong interest in marine life led him to build an aquarium in which he housed marine animals for long periods of time. When he returned to England, he published his experiences in books. One of them was written in the form of conversation with his son (a son he had not yet conceived). Later, he traveled to Jamaica and continued his naturalist activities in his spare time. Upon returning to England, he published more books and became a popular science writer with an engaging literary style. Among his published books, there was one without which he might have been better off. In *Omphalos* (later republished as *Creation*), he communicated a contradictory and confused hypothesis of creation. To his financial and professional embarrassment, this unscientific treatise appeared just 2 years before Darwin's book on natural selection. The sales were disastrous. Afterward, Gosse returned to his more respectable naturalist activities and even corresponded with Darwin. Gosse's blue plaque is a joint one with his son, Sir Edmund Gosse, who became a writer. He wrote about his father in his own book, *Father and Son*.

Robert Fortune (1812–1880) was a botanist famous for his long journeys to China and for bringing back to Britain many different kinds of tea and other plants. During his first journey, he collected and brought back 120 plant species. Later he described his experience in China and with tea in a book and popularized tea drinking. When he went back to China in 1848, it was a different visit with a different purpose. He was acting on behalf of the British East India Company and traveled disguised as a Chinese merchant. He collected not only plants and seed, but even more know-how that the Chinese government had tried to protect by strict enforcement of rigorous regulations. He was like a modern-day industrial spy, and he prevailed. He brought everything to the British colony of India not only seed, plants, and know-how but eight Chinese experts to teach farmers and processors of tea. Fortune's activities contributed a great deal to establishing a strong tea industry in India to the detriment of the Chinese tea industry. Eventually British import of tea from India by far exceeded that from China. This was, of course, not the only way the British crown hurt China; there were also the opium wars, but that is a different story.

Some Fellows of the Royal Society, 1885 (Wellcome Collection). Standing, from left to right: Sir G. H. Darwin, Sir Francis Galton, Sir W. T. Thiselton-Dyer, R. H. Scott, Sir William Huggins, Sir W. H. Preece, Lord Rayleigh, Sir John Evans, Sir E. Ray Lankester, Sir W. H. M. Christie, Sir Edward Frankland, Sir Norman Lockyer, Sir A. W. Williamson; Sitting, from left to right: Sir Gabriel G. Stokes, Sir Joseph Hooker, J. J. Sylvester, Thomas H. Huxley, Sir Archibald Geikie, John Tyndall, Arthur Cayley, Sir Richard Owen, W. H. Flower, Sir William Crookes

This image shows a group of Fellows of the Royal Society from 1885. Three of the biologists discussed in this section are among them, viz., Sir Joseph Hooker (sitting, second from left), Thomas H. Huxley (sitting fourth from left), and Richard Owen (sitting, third from right).

Statue of Thomas Henry Huxley by Onslow Ford, 1900, at the Natural History Museum and his plaque at 38 Marlborough Place, St John's Wood, NW8

Thomas Henry Huxley (1825–1895) was yet another example of a youth who had to give up education early on account of his family's dire financial situation. He was also an example of dedication to acquiring knowledge by self-education. Reading books was a possibility, and learning the German language was also a means for acquiring more knowledge. Huxley apprenticed with a variety of people, among them medical men and college professors. This too was a way of acquiring practical knowledge and skills. He was set on a path of doing research on invertebrates, later also vertebrates, and publishing scholarly papers. He joined the naval expedition of the *Rattlesnake*, 1846–1850, and proved himself an excellent marine naturalist. Despite lacking formal higher degrees, he became a Fellow of the Royal Society and a member of its Council. He interacted with other Fellows, such as Joseph Hooker and John Tyndall, who became his friends. Huxley had a spectacular career in British academia. He was Professor of Natural History at the Royal School of Mines, Fullerian Professor at the Royal Institution, Hunterian Professor at the Royal College of Surgeons, and President of the Royal Society, just to mention some of his functions over the years. His most remarkable title was an informal one that he acquired after he had died: "Darwin's bulldog." In truth, he was a fierce and most knowledgeable defender of the theory of evolution and natural selection. He excelled in public debates, which was the more important as Darwin himself avoided the limelight. Huxley was a prolific author and a great public educator. Some of his offspring rose to fame, among them the biologist Sir Julian Huxley (see below) and the physiologist Sir Andrew Huxley.

Memorial of Frederick DuCane Godman and Osbert Salvin by F. Arnold Wright in the main hall of the Natural History Museum

Frederick DuCane Godman (1834–1919) was an entomologist and ornithologist. Osbert Salvin (1835–1898) was a zoologist and ornithologist. Together, the two have been known especially for their comprehensive studies of the fauna and flora of Central America. They jointly published a 52-volume encyclopedia on the natural history of the continent, the *Biologia Centrali-Americana*, which came out between 1879 and 1915.

Bas-relief honoring the naturalist W. H. Hudson at the Sanctuary at Kensington Park dedicated to his memory. The years of his birth and death are carved into the stone (MDCCCXLI–MCMXXII)

Memorial plaque of the naturalist and writer William Henry Hudson at 40 St Luke's Road, W11, on the occasion of the 150th anniversary of his birth, and his plaque at his former domicile, 14 Leinster Square Terrace, W2

William Henry Hudson (1841–1922) was a British naturalist and ornithologist who lived the first three decades of his life in Argentina. He investigated the local flora and fauna, had a special love of Patagonia, and wrote scholarly papers about his ornithological observations. He settled in London in 1874 and continued his literary activities, both fiction and non-fiction.

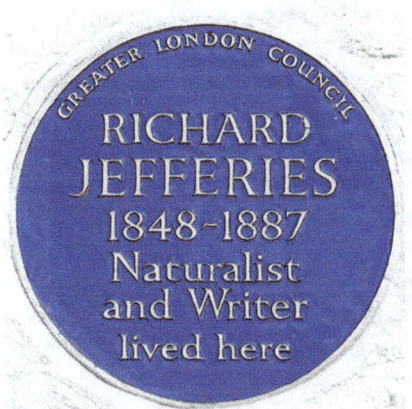

Plaques of Richard Jefferies at 59 Footscray Road, Eltham, SE29; Agnes Arber née Robertson at 9 Elsworthy Terrace, NW3; and Dame Helen Gwynne-Vaughan at Bedford Court Mansions, WC1 (all three, Spudgun67)

Richard Jefferies (1848–1887) was a prolific writer who, in addition to fiction, wrote about natural history for adults and children alike.

Agnes Arber (née Robertson, 1879–1960) was a botanist with her main interest in the morphology of plants. She wrote her first paper in botany for her school's magazine when she was 15. She studied at UCL and at the University of Cambridge and stayed in Cambridge for the rest of her life. She worked for a Laboratory for Women maintained by the University of Cambridge. She was engaged in experimental morphological studies and published three books. When the Laboratory was closed, she continued working at her own laboratory in her house and produced yet another book. From the early 1940s, she stopped bench work and devoted herself to writing. She published works about John Ray, Joseph Banks, and other great botanists of the past. She was fascinated by Goethe's studies in botany and wrote about them extensively. Her main treatise was *The Natural*

Philosophy of Plant Form, published in 1950. She was the third woman elected to be Fellow of the Royal Society (1946) and the first who was awarded the Gold Medal of the Linnean Society (1948).

Dame Helen Gwynne-Vaughan (née Fraser, 1879–1967) was a botanist and mycologist. She was among the first women who could be a regular student at King's College in London. She was interested in taking a degree in botany and/or in zoology, but zoology was considered to be too unladylike; hence, she went on to become a botanist. Her scientific interest focused on the cytology of fungi and she published extensively in this field during the 1920s and 1930s. Numerous papers followed, and she co-authored a book (with B. Barnes), *The Structure and Development of the Fungi* (1927), which then was reprinted several times. Gwynne-Vaughan taught at the Royal Holloway College (today, it is part of the University of London), at UCL, Nottingham, and at Birkbeck College, from which she retired in 1944. She was a pioneer not only in science but also in the military. Both in WWI and in WWII, she served in leading positions of women auxiliary corps. She had an assertive personality when this was even more unusual than today for a woman.

Plaque of three Huxleys: Julian Huxley, his father, and one of his brothers at 16 Bracknell Gardens, NW3 (Spudgun67), and Julian Huxley's alone at 31 Pond Street, NW3

Leonard Huxley (1860–1933) was one of Thomas Huxley's children. He produced three volumes of *Life and Letters of Thomas Henry Huxley*, two volumes of the *Life and Letters of Sir Joseph Dalton Hooker*, a short Darwin biography, and other treatises. Aldous Huxley (1894–1963) was Leonard's son and Julian's brother. He was a writer and a Hollywood screenwriter. His education began in botany at home where Leonard maintained a well-equipped laboratory. Aldous's interests then took him to a different though very successful career. His most famous book is the dystopian novel *Brave New World* (1932).

Julian Huxley (1887–1975) was the grandson of Thomas Huxley and the son of Leonard Huxley. He graduated from the University of Oxford, majoring in biology. He spent some time in Germany and Texas and returned to England during WWI to participate in the war effort. In 1925 he received appointment to King's College as Professor of Zoology, but interrupted his position to take up a project with H. G. Wells and his son to produce a volume *The Science of Life*. Huxley was Fullerian Professor of Physiology at the Royal Institution between 1927 and 1931. He was leading a very active life in which scientific and political projects intermingled. In 1935, he made fun of those German biologists, who were preoccupied with racial purity. He described the "ideal Teuton" as being "as blond as Hitler, as tall as Goebbels, as slim as Goering ...."[15] Huxley traveled to Africa, the United States, and even to the Soviet Union. He took a stand against the unscientific Trofim Lysenko, who was destroying bona fide scientists and the science of genetics in the Soviet Union. Huxley was instrumental in setting up the educational, scientific, and cultural organization of the United Nations, UNESCO, and was its first Secretary General. Huxley excelled in interpreting science for laypersons, and when UNESCO set up an award for the popularization of science, the Kalinga Prize, he received it in 1953, the second year it was given (the first in 1952 went to Louis de Broglie). Huxley was involved in many projects and causes, for example, in the eugenics movement and the protection of wildlife. He was one of the most decorated individuals among scientists.

---

[15] See in Walter Gratzer, *The undergrowth of science. Delusion, self-deception and human frailty* (Oxford: Oxford University Press, 2000), p. 301.

Medicine Men and Women

Statue of William Harvey by Joseph Durham at Burlington House

Left: "Physiology" (upper level, southwest corner, by John Birnie Philip, 1868). Right: "Medicine" (upper level, northeast corner, by Henry Hugh Armstead, 1868) at the Albert Memorial

"Physiology" and "Medicine" are two of the eight allegorical statues of the Albert Memorial. The science of physiology deals with the functioning of living organisms, whereas medicine involves the diagnosis, treatment, and prevention of diseases. The two are distinct disciplinary areas even if they are strongly interwoven. Alfred Nobel preserved this distinction when he named one of his three science prizes Physiology *or* Medicine. The creators of the Albert Memorial also clearly differentiated between the two. While respecting this distinction, our chapter is titled "Medicine" for simplicity.

We present some allegorical images and a Hippocrates bust, followed by a selection of memorials in a few institutions, namely, the Royal College of Physicians, the Royal Society of Medicine, King's College, and Guy's and St Thomas' Hospitals. The two latter are now associated with King's College. Then we survey a collection of memorials to exceptional early women scientists from the times when women's participation in these fields was rare. This is followed by a review of memorials grouped by medical specializations, which we admit is a rather loose concept, because some of the medical professionals could be considered for more than one group. We conclude the chapter with a note on the Wellcome Foundation.

4 Medicine Men and Women

From left to right: Statue of Asclepius and his rod with a snake twined around it at the British Museum; relief of Hygeia feeding the snake (by Eric Aumonier) above the entrance, Princess Royal Nurses Home, Guilford Street, WC1N; and statue of Asclepius with his rod at the Royal Society of Medicine

Left: Statue of Asclepius; the statue also depicts his son, Telesphorus; both at the Royal College of Physicians. Right: Statues of Asclepius and Hygeia at Guy's Hospital

Asclepius is the god of healing and medicine in Greek mythology. He appears with a serpent-entwined rod symbolizing medicine, healing, and sanitation (Rod of Asclepius or Staff of Asclepius). Hygeia was his daughter, the goddess of health and the prevention of sickness. Her name has developed into the English word *hygiene*. Telesphorus was Asclepius's son who symbolized recovery from illness.

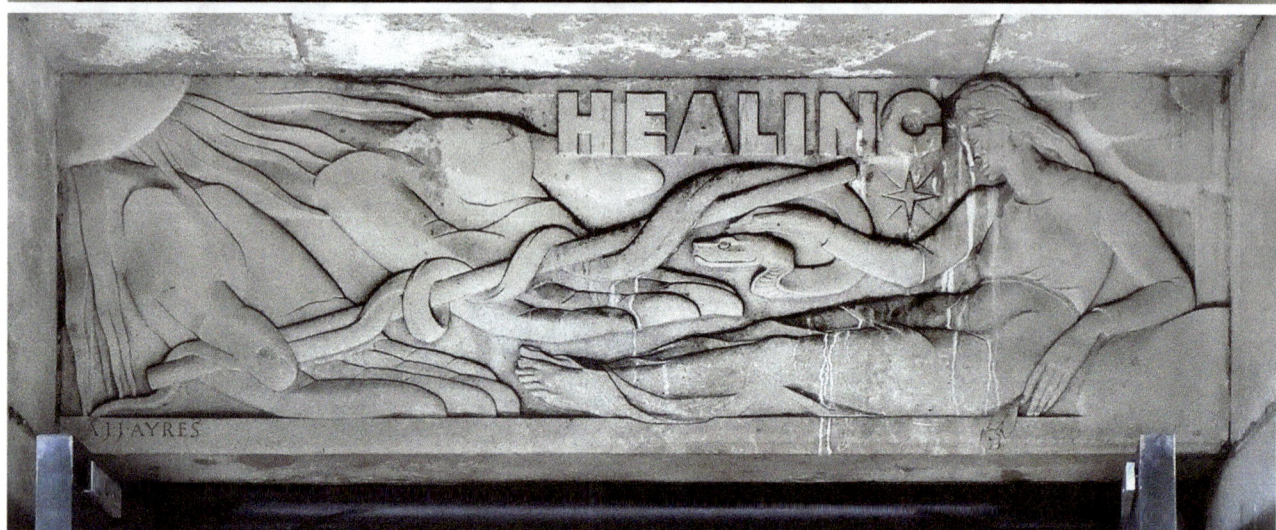

Allegorical representations of Research (top) and Healing (bottom) by A. J. J. Ayres above two of the entrances of the National Hospital for Neurology and Neurosurgery, Queen Square

There are two allegorical relief panels on the façade above two of the entrances to the Queen Mary wing of the National Hospital for Neurology and Neurosurgery. The "Research" panel bears a quotation from the writings of Horace Mann carved on the pages of an open book, "Every addition to true knowledge is an addition to human power." Mann (1796–1859) was an American educational reformer and abolitionist. The historian Ellwood P. Cubberley said of Mann, "No one did more than he to establish in the minds of the American people the conception that education should be universal, non-sectarian, free, and that its aim should be social efficiency, civic virtue, and character … ."[1] At one time the research department of the institution carried the name Institute for the Teaching and Study of Neurology, hence the emphasis of educational aspects on the "Research" panel. The "Healing" panel can be interpreted as if two hands, presumably belonging to God and illuminated by the Sun, offer the Rod of Asclepius to a beautiful young girl, possibly symbolizing humankind.

---

[1] E. P. Cubberley, *Public Education in the United States* (1919), p. 167.

4 Medicine Men and Women

The Hippocrates bust was based on a Hippocrates sculpture at the British Museum. Hippocrates (c. 460–370 BCE) was a Greek physician and philosopher, revered as the "father of medicine." The Hippocratic Oath is a set of ethical rules which used to be sworn to by medical students upon graduation. Today, only a few of the medical schools use its original text; others use a modified version; and some use something else. The bust used to decorate a medical bookstore in the building managed by Henry King Lewis and his company. The building and the bust were erected at the same time.

Hippocrates bust by the Aumonier sculptor firm, 1930, 136 Gower Street, WC1E

 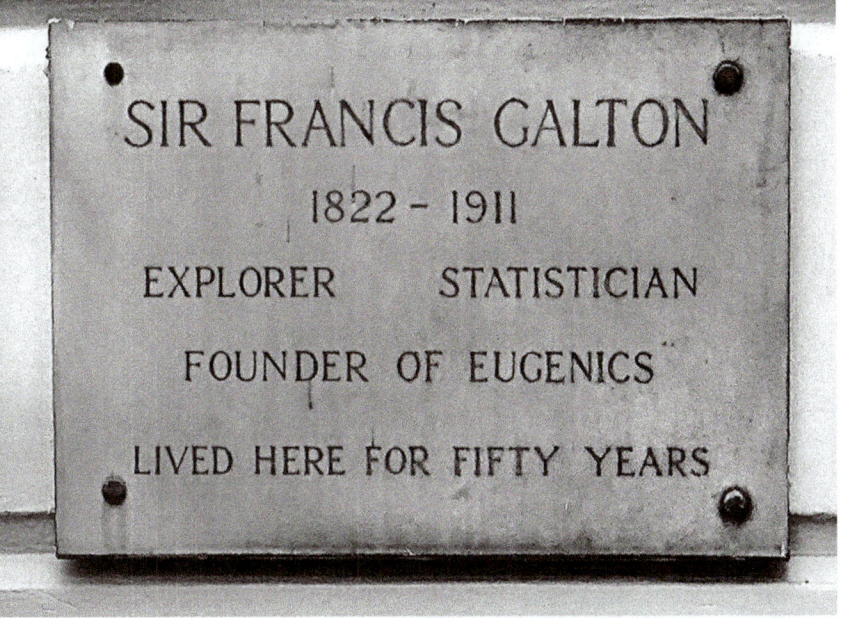

Portrait of Sir Francis Galton (Wellcome Collection) and his plaque on the façade of 42 Rutland Gate, Knightsbridge, SW7

Sir Francis Galton (1822–1911) lived for 50 years in the house where his memorial tablet was erected. It refers to him as explorer, statistician, and the founder of eugenics. It is the latter that has made him both renowned and notorious. He initiated new concepts in statistical analysis and investigated human traits and human communities using questionnaires, which was a pioneering approach in social science. He classified fingerprints. He coined the famous expressions of "eugenics" and popularized the "nature versus nurture" debate. In 1869, he published *Hereditary Genius* in which he described the hereditability of intellectual performance, but ignored the importance of the conditions and circumstances in which children of different classes grow up. He promoted the idea of offering incentives to people to form genetically advantageous marriages. One of the incentives he envisioned would be a wedding ceremony, free of charge, in Westminster Abbey. He studied the efficacy of prayer for enhancing longevity and found it nonexistent. He was an innovator in meteorology by introducing the scientific method in weather forecasting and advocated the utility of record keeping in understanding short-term climate changes. Galton was a polymath with a tremendous output of writing. It is not so much his teachings, rather, their interpretation by his followers, that have become immensely controversial.

Royal College of Physicians

Thomas Linacre's bust by Henry Weekes, 1876, in the garden of the Royal College of Physicians, 11 Saint Andrews Place, Regent's Park, NW1, and his portrait by H. Cook (Wellcome Collection)

Thomas Linacre (c1460–1524) was a physician but is best known as a humanist. His pupils included world-renowned scholars, such as Erasmus, and other notables, such as Sir Thomas More. Linacre was an educator and an intellectual trendsetter and his name is commemorated broadly. He led the physicians in 1518 who petitioned King Henry VIII to establish what has become known as the Royal College of Physicians (RCP). The reason was that the medical profession at that time was poorly regulated. Today, the RCP has some 36,000 members, internationally. It provides its members accreditation by examination.

Left: Bust of William Harvey, 1739, Royal College of Physicians. Right: Painting (ca. 1900) of William Harvey performing autopsy of Thomas Parr; unknown artist (Wellcome Collection)

William Harvey (1578–1657) discovered the circulation of blood. He was the son of a well-to-do farmer, so his family circumstances provided for these studies. He went to school in Canterbury and, at the age of 16, attended the University of Cambridge. He graduated at the age of 19 and continued for 5 years at the University of Padua in Italy, where he became a physician with a special interest in anatomy. He returned to England, settled in London, built up a thriving private practice, and in 1605 married the daughter of a famous physician. One of his patients was Francis Bacon. Harvey became a member of the Royal College of Physicians and professor at the St Bartholomew's Hospital. In 1616 he was honored by a 7-year appointment to the Lumleian Lectureship, which had been established for disseminating knowledge of anatomy throughout England. He added a third to his two functions in 1618, when he became personal physician to James I and later to Charles I. Harvey's treatise about the circulation of blood, *Exercitatio Anatomica de Motu Cordis et Sanguinis in Animalibus* (*Anatomical Study of the Movement of Heart and Blood in Animals*), was published in 1628 in Frankfurt, Germany. Werner Forssmann who received his Nobel Prize in 1956 for heart catheterization called this work in his Nobel lecture the birth of cardiology. Harvey spent 3 years traveling throughout France and Spain, during which he continued his studies of anatomy. Upon his return home, there was much political turmoil in England, but he managed to continue his scholarly life and he received numerous high recognitions for his achievements.

Harvey was involved in a famous inquest about Thomas Parr. This man claimed that he was born in 1483, so he would have been 152 years old when he was invited to meet Charles I. Parr lived in poverty on a most simple diet. According to a legend, he died while being treated to a lavish meal by the King, when Parr choked on some rich food. The gist of this story was that when he was taken out of his natural environs, the sudden change in his diet may have harmed him. The King arranged for Parr to be buried at Westminster Abbey, where his gravestone still is. Its inscription gives 1483 for his birth and 1635 for his death and lists the names of the ten kings under whose reign he supposedly lived. Before Parr was buried, Harvey dissected Parr's cadaver and found his internal organs in perfect shape. Modern estimates give about 70 years for Parr's true age at the time of his death.

In 2000, the noted historian Arthur Schlesinger, Jr., compiled a list of the ten most influential people of the second millennium for the *World Almanac* and included Harvey with nine others: William Shakespeare, Isaac Newton, Charles Darwin, Nicolaus Copernicus, Galileo Galilei, Albert Einstein, Christopher Columbus, Abraham Lincoln, and Johannes Gutenberg. Note that out of the ten, eight belong to the subject matter of this book, viz., scientists, inventors, explorers, and pioneers in medicine. However subjective such a list may be, it gives an estimate of William Harvey's greatness.

From left to right: Portraits of Richard Lower by Jacob Huysmans; William Hewson by unknown artist; and Stephen Hales by J. McArdell (after T. Hudson) (all three, Wellcome Collection)

Harvey's discovery of the circulation of blood was the starting point of the principal achievements of the physicians Richard Lower (1631–1691) and William Hewson (1739–1774), and the clergyman Stephen Hales (1677–1761). Lower studied in London and Oxford and came across many of the luminaries of his age. He knew John Locke, the philosopher. The physician Thomas Willis (1621–1675) was Lower's mentor. Willis was involved in anatomy, neurology, and psychiatry and was one of the founders of the Royal Society (Chap. 3). Lower interacted with the natural scientist Robert Hooke, famous for his microscope and coining the biological term *cell*. Lower investigated the heart and pioneered blood transfusion. He had a rich publishing record in the medical literature and was court physician to Charles II during the King's last illness.

Hewson studied in Newcastle and Edinburgh and at one time was William Hunter's assistant. Hewson investigated the blood and its pathology, and he is considered the "father of hematology."

Hales received degrees in mathematics, among others, in addition to his theological studies in Kensington, Orpington, and Cambridge. He was interested in physiology, chemistry, and botany and made discoveries in all three. He advocated the importance of air quality for living organisms, especially plants. He provided precise information of the capacity of an animal heart and was the first who measured blood pressure in 1727. Hales also carried out the first catheterization of the heart of a living animal.

Busts of Thomas Sydenham by Joseph Wilton, 1758 (left), and William Babington by William Behnes, 1831 (right)

The physician Thomas Sydenham (1624–1689) became known as "the English Hippocrates" due to his most successful text, *Observationes Medicae*. His recognition has been mostly posthumous and he has also been called "the father of English medicine." One of his discoveries was the disease known as Sydenham's chorea or St Vitus' dance, an abnormal involuntary movement disorder.

William Babington (1756–1833) studied medicine at Guy's Hospital and spent a considerable period of his career there. He also taught chemistry, was a noted geologist, and served as president of the Geological Society of London.

From left to right: Bust of Richard Bright by William Behnes, 1871, and his plaque, 11 Savile Row, W1, and bust of John Conolly by Giovanni Maria Benzoni, 1866

The physician Richard Bright (1789–1858) pioneered research on kidney disease. He identified what has become known as Bright's disease and is considered the "father of nephrology." A memorial plaque honors him in Keszthely, Hungary, remembering his visit at and love for Lake Balaton.

John Conolly (1794–1866) was a psychiatrist who established a pioneering non-restraint treatment of the insane. His method was followed throughout England and by many internationally. The renowned Russian psychiatrist Sergey S. Korsakov (1854–1900) became acquainted with Conolly's approach during his travel in France and Germany and adopted it in Moscow. Korsakov's bust stands today in the garden of the Sechenov Medical University in Moscow.

A painting (not shown) at RCP honors Sir Raymond Hoffenberg (1923–2007), the South-African-born endocrinologist whose main research interest was the thyroid. He fought in WWII, worked under Albert Schweitzer, and assisted Christiaan Barnard in the preparations for the first heart transplant in 1967. Hoffenberg opposed apartheid, so he had to leave South Africa in 1968. He continued a most successful career in the United Kingdom. He was active in medical education and in medical organizations and institutions. He was President of the Royal College of Physicians between 1983 and 1989.

Royal Society of Medicine

The entrance to the Royal Society of Medicine (RSM), 1 Wimpole Street, Marylebone, W1G, shown along with the plaques of Sir Henry Wellcome (left) and Sir Henry Dale (right), at the RSM

The Medical and Chirurgical Society of London was established in 1805 for sharing medical and healthcare knowledge. (*Chirurgical*, derived from the same Greek word for hand as *chirality* in chemistry, refers to the origins of surgery as a hand skill.) It was renamed the Royal Medical and Chirurgical Society under the Royal Charter by William IV in 1834. Merging with 15 specialist medical societies in 1907, the Royal Society of Medicine (RSM) of today was established under a supplemental Royal Charter by Edward VII. The present headquarters were inaugurated in 1912. The Society includes many of the city's most famous scientists: Charles Darwin, Edward Jenner, Sigmund Freud, Sir Alexander Fleming, and Thomas Addison (the discoverer of Addison's disease, a deficiency of the adrenal gland from which US President John F. Kennedy suffered). Sir Henry Wellcome was a benefactor and Honorary Fellow of the Society. Sir Henry Dale was an Honorary Fellow and a former President of the Royal Society. There is more about both later in this chapter.

Portraits of Erasmus Darwin, color mezzotint by J. R. Smith, 1797, after J. Wright, and Sir William Osler, oil painting by Harry Herman Salomon after a photograph; both, Wellcome Collection

Erasmus Darwin (1731–1802) was a physician, poet, philosopher, botanist, naturalist, and Charles Darwin's grandfather. He had ideas about evolution and advocated an "integrated approach" to the progress of knowledge.

Sir William Osler (1849–1919) was a Canadian physician. His acumen in his profession is manifested in the many diseases, conditions, and symptoms that carry his name. He was one of the four professors of medicine who founded the Johns Hopkins Hospital in Baltimore. He was an influential author and published his most famous book, *The Principles and Practice of Medicine*, in 1892, which has remained in print (its latest reprint appeared in 2018). This textbook played a role in establishing the Rockefeller Institute (today, University). John D. Rockefeller, who amassed enormous wealth, was looking for a proper charity, and the Baptist minister, Frederick Taylor Gates (1853–1929), was his advisor. Gates read Osler's text from which he realized that there were so many instances when it was impossible to properly describe medical conditions, let alone treat them. Gates persuaded Rockefeller to found an institute for medical research, and the Rockefeller Institute (as it was then) opened in 1901.

There are several busts in the RSM Library. Three of them stand rather inconspicuously under the stairs to the Gallery.

Busts under the stairs to the Gallery at the RSM Library. From left to right: James Marion Sims by Paul Duboy, Sir George Henry Makins, and Edwin Rodney Smith—Lord Smith of Marlow—by Nigel Boonham

James Marion Sims (1813–1883) was an internationally renowned American surgeon who founded in New York City the first women's hospital in the United States in 1855. He was a pioneer of modern gynecology and obstetrics. He spent a few years of his career in London. Recently, there has been increasing criticism of his discriminatory medical practices. He treated his slave patients as sub-humans and carried out unethical human experimentation. His memorial used to stand at the perimeter of Central Park where Fifth Avenue and 103rd Street meet. However, New York City has removed his statue from the pedestal, and there are plans to move it to the cemetery in Brooklyn where Sims is buried.

Sir George Makins (1853–1933) served as consulting surgeon in the Boer War and in WWI. In 1917, he was elected President of the Royal College of Surgeons.

Lord Smith of Marlow (1914–1998) served as a surgical specialist in WWII. After the war, he worked at the St George's Hospital. He was President of the RSM in 1978–1980; prior to that, he was President of the Royal College of Surgeons.

There are busts and an engraved glass screen in the Reading Room. The engraving (not shown here) depicts the portrait of the dermatologist Robert Willan (1757–1812) who classified skin diseases (see below).

Three busts in the Reading Room, from left to right: John Elliotson, Sir James Young Simpson, and Robert Barnes (this third one by Hamilton P. MacCarthy)

John Elliotson (1791–1868) did pioneering research into the nature of allergy. He had senior positions at UCL and at University College Hospital. He was a prolific author and an excellent clinician. Eventually, he had to give up his positions because of his entanglement in phrenology and mesmerism, and he founded the London Mesmeric Infirmary. Thomas Wakley (see later in this chapter), the founder of *The Lancet*, fought relentlessly to oppose Elliotson's pseudoscientific views.

Sir James Young Simpson (1811–1870) was an obstetrician, who first demonstrated the use of anesthetics in childbirth: first, nitrous oxide, then, chloroform. According to a myth, to mark the first use of chloroform during childbirth, the parents of the newly born named their baby Anaesthesia. In reality, it was Simpson who gave the baby this nickname.

Robert Barnes (1817–1907) was an obstetrician, in his time the leading specialist in London. He initiated the British Gynecological Society.

King's College

A series of large photographs of scientists are displayed along the Strand and Kingsway on the downtown campus of King's College. These scientists had been associated with the College at one time or another, for longer or shorter times. Here we present a few of those belonging to the health professions and not figuring elsewhere.

James Africanus Horton and Sir Kelsey Fry

Sir James Black, Anthony Clare, and Dame Nancy Rothwell

James Africanus Horton (1835–1883) was from Sierra Leone. He studied medicine at King's College and in Edinburgh. Later serving as a medical officer in West Africa, he was a champion of independent African political thought.

Sir Kelsey Fry (1889–1963) was a dentist and a pioneer in facial reconstruction. During WWII he worked in maxillofacial and plastic reconstruction for the military, among them the Royal Air Force. The treatment of injuries and damages to the jaw bridged the competence of surgeons and dentists. Fry developed this area of medicine.

Sir James Black (1924–2010) was a pharmacologist who developed the beta blocker propranolol for the treatment of heart disease and the H2 receptor antagonist cimetidine for the treatment of stomach ulcers. In 1988, he was co-recipient of the Nobel Prize in Physiology or Medicine. The formulation of the motivation for the prize referred to the discoveries of important principles for drug treatment rather than to specific drugs.

Anthony Clare (1943–2007) was a psychiatrist. He studied at University College Dublin and trained at the Institute of Psychiatry, which later became part of King's College. In addition to his doctorate in medicine, he held a master's degree in philosophy. In the last decades of the twentieth century, he was the most famous psychiatrist in Britain. He directed institutions and authored popular books in psychiatry. He was a prominent broadcaster of popular science programs of the BBC.

Dame Nancy Rothwell (1955–), a physiologist, has held leadership positions both in academia and in the pharmaceutical industry. She studied at the University of London and in

1987 she was awarded a Doctor of Science degree by King's College. Her research interests have included obesity and involuntary weight loss, the role of inflammation in brain disease, and the treatment of stroke.

St Thomas' Hospital

St Thomas' Hospital, Westminster Bridge Road, SE1, is a teaching hospital and a member of King's College School of Medical Education, just like Guy's Hospital (see below). Its name needs some clarification. Originally, it was the Hospital of St Thomas, named after the martyr Thomas à Becket (1118–1170). Henry VIII had it closed in 1540 and Edward VI re-founded it in 1551 as the Hospital of St Thomas', named after the Apostle St Thomas.

Statue of Mary Seacole by Martin Jennings, 2016, in front of St Thomas' Hospital

There are two large sculptures in front of St Thomas' Hospital. One commemorates Mary Seacole (1805–1881), the first memorial to a black woman in the United Kingdom. She was a British-Jamaican nurse who demonstrated self-sacrifice and dedication during the Crimean War although her services were declined by the authorities. She attended and comforted the wounded and applied Jamaican and West African remedies to those in her care. She received much recognition posthumously although there is still controversy between her supporters and those of Florence Nightingale. There is more about Mary Seacole later in this chapter.

Sculpture "Cross the Divide" by Rick Kirby, 2000, on the approach bridge, from Westminster Bridge Road to the main entrance of St Thomas' Hospital.

As for the second large art piece in front of St Thomas' Hospital, "Cross the Divide," "the sculpture represents a helping hand and also trust and relationship on several levels – between patient and healer, the NHS Trust and the medical team, and the joining of Guy's and St Thomas'," according to the sculptor.[2]

---

[2] From an interview by Neeta Borah, January 2008, http://rickkirby.com/interview.htm (downloaded October 5, 2019). NHS is the National Health Service.

St Thomas' Hospital

Statues of King Edward VI by Thomas Cartwright, 1682 (left), and by Peter Scheemakers, 1737 (right)

There are two statues of Edward VI, the re-founder of the hospital. One is immediately at the entrance and the other on the way from the entrance to the Central Hall. Edward VI (1537–1553) was the son of Henry VIII and ascended to the throne in 1547, when he was 9 years old. He re-founded St Thomas' Hospital when he was 14 years old.

The Central Hall of St Thomas' Hospital is a busy place, being a comfortable venue for students and visitors sitting on benches. There is a grand piano at the end of the Hall in front of a statue of Florence Nightingale. This Nightingale statue of bronze resin is a replica of the one of bronze that was stolen, which itself was a reduced replica of Arthur George Walker's 1915 Nightingale statue of the Crimean War Memorial at Waterloo Place. The original statue is shown below under the section Nurses.

Two busts flanking the Nightingale statue in the Central Hall of St Thomas' Hospital. Facing the statue, on the left, Dame Cicely Saunders by Shenda Amery (2002) and, on the right, Theodora Turner by Robert Dawson (2002)

Dame Cicely Saunders (1918–2005) was the founder of the modern hospice movement. She was trained as a medical social worker, and in 1957 she qualified as a doctor, which added weight to her teachings. She researched the issue of terminal care and was a prolific author on this subject. In 1989 she was named to the Order of Merit (OM), the pinnacle of the British honors system.

Theodora Turner (1907–1999) was trained as a nurse and performed her duties both in peacetime and in war. In WWII, she was there at the evacuation of the British troops in Dunkirk and participated in rebuilding St Thomas' Hospital after 13 hits by German bombs. At the end of her career, she was the Matron of Nurses at St Thomas' Hospital.

The two longer sides of the oblong, rectangular-shaped Central Hall are lined with five busts each. These busts of physicians and surgeons are presented here with increasing order of birth years.

Bust of Richard Mead by Henry Weekes, 1871, and Mead's portrait—mezzotint by R. Houston after A. Ramsay (Wellcome Collection)

Richard Mead (1673–1754) was a forward-looking physician of his time whose special interest was preventive medicine. He wrote about the means of preventing and treating plague, smallpox, measles, and scurvy. He was physician to Royalty, leading politicians, and Isaac Newton.

Bust of William Cheselden by Henry Weekes, 1871, and a painting of Cheselden giving an anatomical demonstration, attributed to Charles Phillips (Wellcome Collection)

William Cheselden (1688–1752) studied medicine at St Thomas' Hospital. He became a leading surgeon and a lecturer in anatomy and surgery with special interest in the anatomy and pathology of the bones. He pioneered new approaches in extracting bladder stones and in alleviating the burden of blindness. For his achievements in surgery, he was appointed to be Queen Caroline's surgeon. He played a leading role in the movement to separate the surgeons from the Company of Barbers and Surgeons. The new organization, called the Company of Surgeons, created its own anatomy theater. The Company was the forerunner of the Royal College of Surgeons of England.

Busts of William Lister by William Behnes, 1847 (left), and John Flint South by Henry Weekes, 1872 (right)

William Lister (1757–1830) was a physician at St Thomas' Hospital and later in his career he was in charge of the Hospital administration.

John Flint South (1797–1882) received his training at St Thomas' Hospital and he stayed with the Hospital. He was a surgeon and a lecturer of anatomy with additional appointments as a surgeon elsewhere. He wrote popular treatises about household surgery and handbooks for emergencies.

Busts of Samuel Solly by William Day Keyworth the Elder, 1867, and of Frederick Le Gros Clark by George Gammon Adams, 1855

Samuel Solly (1805–1871) was a surgeon at St Thomas' Hospital. His research focused on the human brain. He gave lectures in anatomy and physiology.

Frederick Le Gros Clark (1811–1892) was a surgeon at St Thomas' Hospital. He was a lecturer in anatomy and physiology. Later, he held the St Thomas' Chair of Surgery.

Bust of Sir John Simon by Frank Theed, 1906, and his plaque, 40 Kensington Square, W8 (Spudgun67)

Sir John Simon (1816–1904) was a physician, pathologist, and an influential public health reformer. He studied at King's College and St Thomas' Hospital. When the Public Health Act was passed in 1848, the General Board of Health was created, and Simon was appointed the first Medical Officer of Health for London. This was the second such appointment in Britain; the first was in Liverpool. In this capacity, Simon accomplished a great deal in improving public health and the hygiene of urban life.

Busts of John Syer Bristowe by unknown sculptor (left) and Charles Murchison by Edwin Roscoe Mullins, 1880 (right)

John Syer Bristowe (1827–1895) received his training at King's College and was a physician at St Thomas' Hospital. He was also appointed to be a pathologist and the curator of the Hospital's museum. He had broad interests and lectured in the medical school in botany, anatomy and physiology, and pathology. When a debate developed about the new location of St Thomas' Hospital, he and Timothy Holmes of St George's Hospital published a report on hospitals in Britain. The principal antagonists in the debate were Sir John Simon and Florence Nightingale. Simon wanted the Hospital to stay in London, but Nightingale objected that the location on the banks of the Thames was unhealthy. Bristowe and Holmes's analysis sided with Simon's preference. They showed that the hospital's siting was less important than the condition of the hospital.

Charles Murchison (1830–1879) was a physician at St Thomas' Hospital and a lecturer at its medical school. When in 1873 an epidemic of typhoid fever broke out in west London, he determined that polluted milk was the culprit. He was a prolific author of medical treatises.

Bust of Sir William MacCormac by Alfred Drury, 1903, and photograph of Sir William MacCormac about to perform a surgical operation at the Bellevue Hospital, New York. Albumen print, 1891 (Wellcome Collection)

Sir William MacCormac (1836–1901) was an internationally renowned surgeon at St Thomas' Hospital who gave lectures at its medical school. He was appointed surgeon to the Prince of Wales, later, Edward VII. One measure of his fame was that Professor Lewis A. Sayre of the Bellevue Hospital in New York, one of the pioneers of orthopedic surgery, asked MacCormac to give a clinical lecture in the amphitheater of the Bellevue and to perform an operation in front of an assembled audience.

Guy's Hospital

Guy's Hospital, St Thomas' Street, SE1, was founded in 1721 by the philanthropist Thomas Guy (1644–1724) for treating the "incurable." Today it is a teaching hospital and a member of the King's College School of Medical Education.

Left: Statue of Thomas Guy by Peter Scheemakers in the forecourt of Guy's Hospital. Right: Memorial tablet of Thomas Guy's birthplace at Tower Bridge Road, SE1, close to his actual birthplace at 7 Pritchard Alley. The Latin inscription *DARE QUAM ACCIPERE* means "giving [is more blessed] than receiving" (this image, Matt Brown)

The original buildings of Guy's Hospital face St Thomas' Street. Statues of Asclepius and Hygeia that decorate the façade are shown in the introductory section of this chapter. The original buildings of the Hospital form a courtyard with the statue of Thomas Guy in its middle, surrounded by an iron railing.

Statues of John Keats (left, by Stuart Williamson, 2007) and Lord Nuffield (right, by Maurice Lambert, 1944) in their respective quadrangles at Guy's Hospital

There are two quadrangles on the campus and each has a statue. One of them honors the Romantic poet John Keats (1795–1821). Keats enrolled as a medical student at Guy's Hospital in 1815 and for a while it seemed that he had found his calling. Already in 1816 he was granted his apothecary's license, which qualified him to practice as an apothecary, physician, and surgeon. Increasingly though, his dedication to poetry overtook his medical interests, so he left medicine to the greater benefit of our universal culture. The other quadrangle is dedicated to Lord Nuffield (William Morris, founder of Morris Motors, 1877–1963), a benefactor and chairman of Guy's Hospital. His statue was erected while Lord Nuffield was still alive.

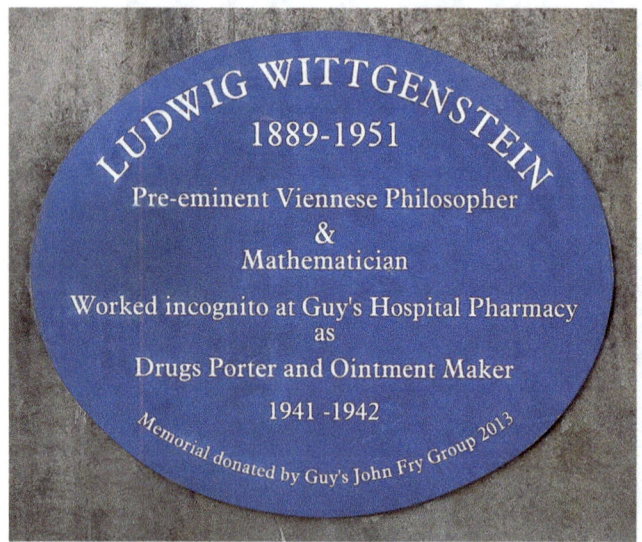 

Memorial plaques of Ludwig Wittgenstein and John Fry under the arcades

Ludwig Wittgenstein (1889–1951) is another polymath outside the realm of medicine who is commemorated on the Guy's campus. His modest memorial plaque says that the "pre-eminent Viennese philosopher and mathematician worked incognito at Guy's Hospital Pharmacy as drugs porter and ointment maker 1941–1942." Wittgenstein was one of the most influential philosophers of the twentieth century, mostly for his works published posthumously. He taught at the University of Cambridge 1929–1947, but during the war he wanted to get involved with manual work, hence his job at Guy's. He befriended some young doctors interested in philosophy and had a gallstone removal operation at the Hospital.

Another memorial plaque, next to Wittgenstein's, honors John Fry (1922–1994) who qualified at Guy's Hospital in 1944 and had a remarkable career as a general practitioner. Fry published extensively on primary care, taught in a number of countries, and was appointed a Fellow of the Rockefeller Foundation. He was a founding member of the Royal College of General Practitioners and was active in a number of medical organizations and institutions.

Frederick Newland Pedley (1855–1944), a physician and dentist, is honored with a simple tablet (not shown). He founded the Dental School of Guy's in 1889. The tablet was erected in 1994, the 50th anniversary of his death.

Keats House, 24–26 St Thomas' Street, designed by Newman and Billing and built in 1863, and portrait of Sir Astley Paston Cooper (Wellcome Collection). The four busts by J. W. Seale in the arches above the windows of the first floor represent, from left to right, William Harvey, Thomas Guy or John Keats, Hippocrates, and, possibly, Sir Astley Paston Cooper

Keats House used to be a part of Guy's Hospital; today it is a clinic of private psychiatrists. There is some uncertainty as to the identity of the four bust reliefs on its façade. Of them, Sir Astley Paston Cooper (1768–1841) has not figured yet in our discussion. He was a surgeon and innovator in medical procedures, especially vascular surgery. He described diseases heretofore not identified that then were named after him, such as Cooper's neuralgia, Cooper's disease (benign cysts of the breast), and Cooper's hernia. He was a meticulous anatomist who identified and described several anatomical structures for the first time, including Cooper's ligaments and Cooper's fascia.

London School of Hygiene & Tropical Medicine

The London School of Hygiene & Tropical Medicine, 2 Keppel Street, the corner of Keppel and Malet Streets

Sir Patrick Manson, from left to right, memorial tablet with relief portrait in the entrance lobby; memorial plaque on the façade of his former domicile, 50 Welbeck Street, W1; and portrait by Harry H. Salomon (Wellcome Collection)

London School of Hygiene & Tropical Medicine

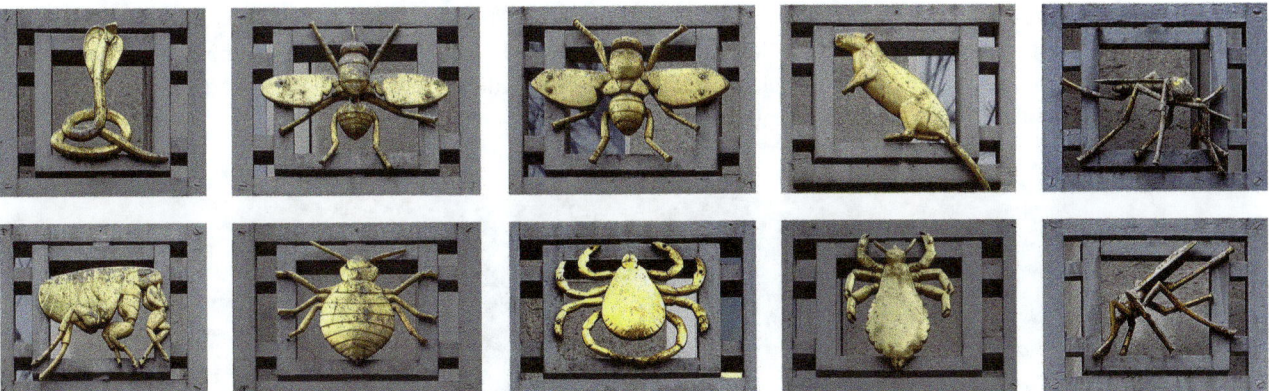

Gilded disease vectors decorating the balconies of the Keppel Street façade. Top row: Cobra, tsetse fly, housefly, black (or brown) rat, Aedes mosquito. Bottom row: Flea, house bug, tick, louse, Anopheles mosquito

The London School of Hygiene & Tropical Medicine was founded in 1899 and is one of the colleges of the University of London. It has become a world leader of education and research in its field. The Aberdeen-trained physician Sir Patrick Manson (1844–1922) founded the School. With his extensive research in parasitology, he laid the foundation for a new branch of medicine, becoming known as the "Father of Tropical Medicine." The current building of the School opened in 1929. Manson's name is one of the 23 contributors to public health and tropical medicine carved in the frieze of the building. A few others are mentioned elsewhere in this chapter: Sir Edwin Chadwick, Edward Jenner, Sir Joseph Lister, Sir John Simon, and Thomas Sydenham. The rest of the names are listed below in alphabetical order.

Hermann Michael Biggs (1859–1923) was an American physician who applied bacteriology to the prevention and control of infectious diseases.

From left to right: Portraits of Sir David Bruce (British Council.), Sir Edwin Chadwick (both, Wellcome Collection), and William Farr (Wikipedia)

Sir David Bruce (1855–1931), a pathologist and microbiologist, investigated Malta fever. It is a severe condition first diagnosed among the British military stationed on Malta following the Crimean War. Bruce found the organism responsible for this condition, which has become known as *brucellosis*.

William Farr (1807–1883), epidemiologist, studied morbidity and mortality and used statistics in his studies. His activities were at the time of heavy epidemics in London. He amassed enormous amounts of data, and this was a constructive aspect of his activities, which were also turning attention to environmental issues. His trustworthy statistical data helped John Snow's research of the possibility that the epidemics were caused by contaminated water. Nonetheless, Farr subscribed to the then-popular, but mistaken, theory according to which cholera, typhus, and the plague were all caused by "bad air" being released by rotting organic matter. It was only in the 1866 London cholera epidemic when Snow's evidence finally convinced Farr that the culprit was indeed contaminated water.

From left to right: Johann Peter Frank, engraving by Ambroise Tardieu (Wellcome Collection); William C. Gorgas (US National Library of Medicine); and Robert Koch (Wellcome Collection)

Johann Peter [Jean Pierre] Frank (1745–1821) was a German physician and hygienist who summarized his teachings about public health in a nine-volume treatise. His concerns included public sanitation, water supply, sexual hygiene, and food safety.

William C. Gorgas (1854–1920) was a high-ranking US Army surgeon. During the construction of the Panama Canal, he introduced measures that enabled the project to keep the mosquito danger under control and thus fought successfully the yellow fever and malaria scourges.

Robert Koch (1843–1910), a German physician and microbiologist, was a great bacteriologist. He identified the agents of tuberculosis, cholera, and anthrax. He received the Nobel Prize in Physiology or Medicine in 1905 for the discoveries related to tuberculosis.

From left to right: Charles Louis Alphonse Laveran; Sir William Boog Leishman, photograph by Bassano Ltd.; and Timothy R. Lewis (all three, Welcome Collection)

Charles L. A. Laveran (1845–1922), French physician, discovered the parasite causing malaria. This discovery was pivotal to understanding not only the nature of malaria but of all protozoal diseases. He also discovered *trypanosome*, which caused the African sleeping sickness. He received the Nobel Prize in Physiology or Medicine in 1907 for his work in understanding the role of protozoa in causing diseases.

Sir William Boog Leishman (1865–1926) was a pathologist and army physician (namesake of leishmania and leishmaniasis).

Timothy R. Lewis (1841–1886), surgeon, researched tropical medicine in India.

From left to right: James Lind, physician at Haslar; Edmund A. Parkes, photograph by Barraud; and Louis Pasteur, photograph by Nadar (all three, Wellcome Collection).

James Lind (1716–1794), physician, founded naval hygiene in the Royal Navy.

Edmund A. Parkes (1819–1876), physician, was a hygienist for the military.

Louis Pasteur (1822–1895), French chemist and microbiologist, was one of the founders of medical microbiology and one of the greatest scientists of all time.

From left to right: Max von Pettenkofer, wood engraving after J. Kriehuber; Sir John Pringle, oil painting (both, Wellcome Collection); and Walter Reed (US National Library of Medicine)

Max von Pettenkofer (1818–1901), German chemist, pioneered measures for hygiene.

Sir John Pringle (1701–1782), physician, advocated the importance of postmortem putrefactive indication in the occurrence of diseases.

Walter Reed (1851–1902), US Army physician, determined that a mosquito transmits yellow fever.

Sir Ronald Ross (Wellcome Collection) and his plaque, 18 Cavendish Square, W1

Sir Ronald Ross (1857–1932) was a British physician, born in India and trained at St Bartholomew's Hospital, the Royal College of Physicians, the Royal College of Surgeons, and elsewhere. Patrick Mason was one of his mentors. From his youth, Ross's main interest was in music and literature. Still, under the influence of a dedicated teacher, Emanuel E. Klein, Ross became a bacteriologist. Seminal discoveries followed. Ross detected the malarial parasite in the mosquito's gastrointestinal tract. He demonstrated that the mosquitoes carried out infection by injecting the bacterium in anticoagulants deposited in their bite. He was awarded the Nobel Prize in Physiology or Medicine in 1902 for uncovering the etiology of malaria.[3] There was some controversy in that he alone was chosen for this distinction and another scientist, the Italian Giovanni Battista Grassi, was left out. The protesters argued that Grassi and his associates had described the development of malaria in a more complete way than Ross. In fact, the Royal Society had awarded its Darwin Medal to Grassi in 1896. After the Nobel Prize, Ross continued his research on tropical diseases. He was the director of the Ross Institute and Hospital for Tropical Diseases, which was eventually incorporated into what today is the London School of Hygiene & Tropical Medicine.

Lemuel Shattuck (1793–1859), Boston politician, dedicated much of his activities to the improvement of public health.

Left: Memorial tablet with relief portrait of Sir Andrew Balfour in the entrance lobby. Right: Memorial plaque at 25 Gordon Street on the former building of the London School of Tropical Medicine and the Hospital for Tropical Diseases (currently the Department of Mathematics of University College London)

Sir Andrew Balfour (1873–1931) specialized in tropical medicine. He studied in Edinburgh and Cambridge and spent some time in Strasbourg before returning for his doctoral studies in Cambridge and Edinburgh. He earned a degree in public health as well and he studied typhoid fever. When he was in South Africa, Patrick Manson mentored him. Balfour devoted himself to tropical medicine for the rest of his career, including 12 years of service in Khartoum, Sudan. He was appointed director of the London School of Hygiene & Tropical Medicine, and it was under his tenure that the Keppel Street building was constructed.

---

[3] In 2015, the Chinese woman Youyou Tu (1930–) received half of the Nobel Prize in Physiology or Medicine for her discoveries of novel therapy against malaria based upon a notation in an ancient Chinese medical text.

 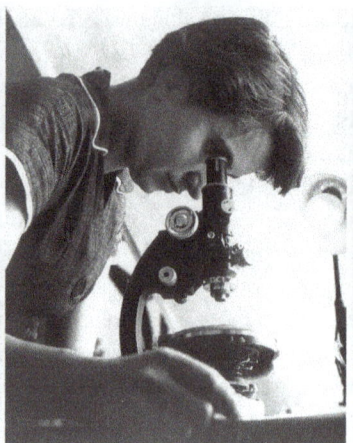 

From left to right: Max Theiler, photograph by Pach Bros., 1951 (Wellcome Collection); Rosalind Franklin (courtesy of Aaron Klug); and James Lovelock in 2005 (photograph by Bruni Comby, Wikimedia)

We mention three names of famous people who had some connection with the London School. Max Theiler (1899–1972) was a South-African-American physician who specialized in virology. He was educated in South Africa and did postgraduate work in London, in St Thomas' Hospital, King's College, and the London School of Hygiene & Tropical Medicine. He then spent the rest of his career in the United States. He received the Nobel Prize in Physiology or Medicine in 1951 "for his discoveries concerning yellow fever and how to combat it."

Rosalind Franklin (see also in Chap. 3), while studying the structure of the polio virus at Birkbeck College in 1957, wanted to investigate the live virus rather than the dead one. This raised concern among her colleagues at Birkbeck. At this point, J. Desmond Bernal asked for help at the London School of Hygiene & Tropical Medicine, which offered to store the live virus safely in their facilities equipped for such tasks. It was a convenient solution as the School and Birkbeck are situated on two sides of Malet Street.

James Ephraim Lovelock (1919–) is an environmentalist, most famous for his Gaia hypothesis. He studied at Birkbeck College and elsewhere and received his PhD degree from the London School of Hygiene & Tropical Medicine.

Women in Medicine

Under ideal circumstances, there should not be a separate chapter for women in medicine. However, most of the period covered by the memorials discussed in this book was characterized with unfavorable circumstance for women in academia. Hence we give added emphasis to the memorials that acknowledge the achievements of women. Not all are mentioned in this section and a few figure elsewhere in this chapter.

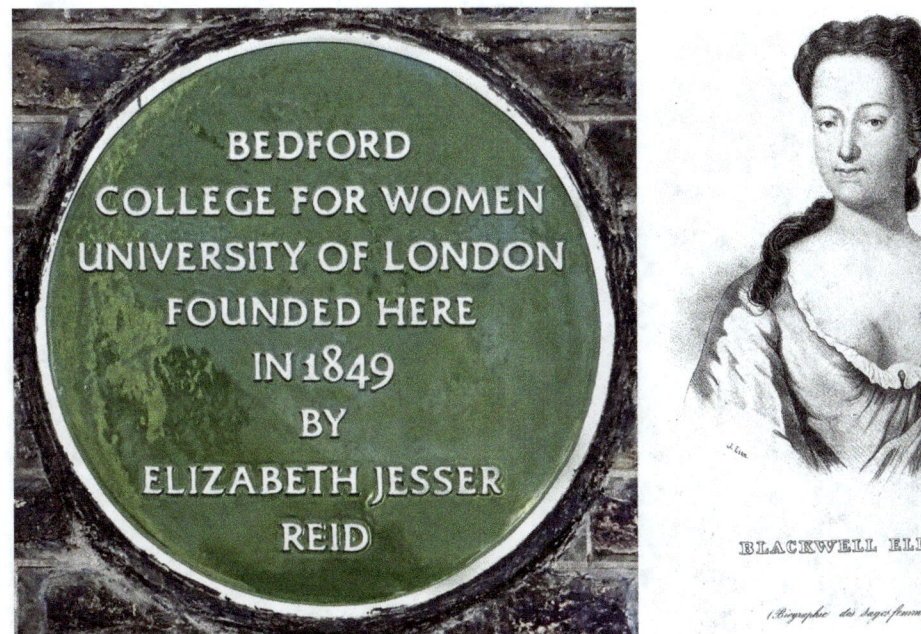

Green plaque of the Bedford College of Women founded in 1849, 48 Bedford Square, WC1, and Elizabeth Blackwell (Wellcome Collection)

The Bedford College of the University of London has undergone considerable transformations since its foundation in 1849. Today it is a coeducational institution under the name of Royal Holloway and Bedford New College, colloquially known as the Royal Holloway, University of London. Elizabeth Blackwell (1821–1910) was one of its notable alumni. She was the first female awarded a medical degree in the United States and entered in the Medical Register of the British General Medical Council. She was born in Bristol, but her family moved to the United States in 1832. She began her career with a variety of teaching jobs. In 1847, she embarked on pursuing a medical degree, still in the United States, and graduated in 1849. In the same year, she returned to Europe and continued her studies in France. She hoped to be accepted as a physician in training as France was considered more progressive than Britain. She was preparing to become an obstetrician though the institution that accepted her as a student insisted to treat her as a student midwife rather than a student physician. Eventually, she moved to London and enrolled at St Bartholomew's Hospital. In 1851, she returned to America and established her practice in New York City. Two years later, she, with two female colleagues, established the New York Infirmary for Indigent Women and Children. In 1858 she returned to England as a new medical act recognized doctors with foreign degrees, and she entered the medical register in 1859. She spent most of the rest of her career in Britain. She had extensive interactions both in the United States and in Britain, mentored Elizabeth Garrett Anderson and others in their medical studies, corresponded with Lady Byron about women's rights, and befriended Florence Nightingale. There are beautiful Elizabeth Blackwell memorials in Asheville, North Carolina, and on the campus of the former Geneva Medical College (now, the Hobart and William Smith College, Geneva, New York). This was the school that admitted her in 1847 after she had been refused by 16 medical schools.

Plaque of Elizabeth Garrett Anderson, 20 Upper Berkeley Street, W1, and her photograph (Wellcome Collection)

If Elizabeth Blackwell is counted as American, then Elizabeth Garrett Anderson (1836–1917) was the first Englishwoman medical doctor. Elizabeth Garrett was born in East London. It was Elizabeth Blackwell who convinced the younger Elizabeth to become a doctor. She studied to become a nurse, but attended classes for male medical doctors until she was barred from doing so on account of the protesting male students. However, she received a certificate from the Society of Apothecaries in 1865 as women were not explicitly excluded by the rules of the Society. Soon afterward, the Society changed its regulations lest other women might go this way in order to become medical professionals.

With her certificate, Elizabeth Garrett opened a dispensary for women in London, became a visiting physician to the East London Hospital in 1870, and founded the New Hospital for Women in London in 1872, staffed exclusively by women doctors (today, it is the Elizabeth Garrett Anderson Hospital for Women). In 1871 she married James Anderson and they soon had three children. She continued her career in medicine as the legislative environment was slowly improving for women entering the medical profession, partially by her example. In 1873, she was accepted as a member of the British Medical Association and remained its only female member for the next 19 years. She founded the London School of Medicine for Women, and in 1883 she was appointed its dean.

The blue plaque shown above decorates the house where Elizabeth Garrett Anderson lived after she married. There is another plaque (not shown) at the site where the house in which she was born stood. Today, the Metropolitan University stands there, 41–47 Commercial Road, E1. There is the Elizabeth Garrett Anderson Wing of UCL Hospital, and an Elizabeth Garrett Anderson Gallery, at 130 Euston Road, NW1.

Women in Medicine

Left: Annie McCall's mosaic portrait on the wall of Morley College for adult education, 61 Westminster Bridge Road, Lambeth, SE1, and her photograph by Deneulain (Wellcome Collection). Right: Jane Harriet Walker (Wellcome Collection)

Annie McCall (1859–1949) was a native of Manchester but studied in several European universities, in Göttingen, Paris, Bern, and Vienna. Then, she continued in the London School of Medicine for Women, the first British medical institution to train women to become doctors. McCall was among its first 50 graduates. She specialized in midwifery and founded the Clapham Maternity Hospital, which was later renamed the Annie McCall Maternity Hospital.

Jane Harriet Walker (1859–1938) specialized in treating tuberculosis. She adopted an approach initiated in Germany whereby the patients were encouraged to spend as much time as they could in the open air rather than in stuffy, warm, indoor environments. She opened sanatoria for treating tuberculosis patients and was a founding member of the Medical Women's Federation and its first president. She was the first woman member of the Council of the Royal Society of Medicine.

Bust of Elsie Inglis by Ivan Mestrovic, 1918, at the Imperial War Museum, Lambeth Road, SE1, and her portrait (Wellcome Collection)

Elsie Inglis (1864–1917) had an enlightened father who supported her determination to study medicine at a time when this was not an accepted career for women. Later in her life, in order to help others, she and her father established the Edinburgh College of Medicine for Women. Eventually she gained her medical qualifications and for some time worked at the New Hospital for Women pioneered by Elizabeth Garrett Anderson. When the University of Edinburgh opened its medical training to women, Inglis obtained her Bachelor of Medicine/Bachelor of Surgery degree in 1899. When in WWI she offered her services to the British War Office, she was refused and in not a very polite way. In contrast, the French Government was happy to take up her offer and she and her team were given an assignment in Serbia. There, she and her colleagues treated wounded Allied soldiers and helped improve hygiene, which reduced the spread of typhus and other infectious diseases. She tried to organize similar efforts for the Russian front, but her cancer prevented her from further actions. In addition to British recognition, she was awarded the highest order of Serbia where she has become a legend.

Two views of the memorial of Dame Louisa Brandbeth Aldrich-Blake in Tavistock Square, designed by Edwin Lutyens. It consists of two identical busts by Arthur George Walker

Dame Louisa Brandbeth Aldrich-Blake (1865–1925) lived decades later than Elizabeth Blackwell, yet she was still among the first British women in medicine. She graduated from the Royal Free Hospital's School of Medicine for Women in 1894. She then continued her studies for a Master in Surgery degree, thereby becoming the first British woman surgeon. She was a pioneer in treating cervical and rectal cancers and had publications in this area of medicine. She worked at her alma mater, devoted much attention to training her students, and from 1914, she was Dean of the School of Medicine for Women. She also took an active role in the British Medical Association. Considering the careers of Elizabeth Blackwell, Elizabeth Garrett Anderson, and Louisa Aldrich-Blake, they contributed to advancing the opportunities of women in medicine, much beyond forging their own advancement. Looking back, the pace of progress may appear slow, but from their perspective, they took giant steps ahead.

A simple tablet "Christine Murrell Memorial" at the corner of Broadley and Lisson Streets (not shown) honors the doctor and psychologist Christine Murrell (1874–1933), the first woman elected to the Council of the British Medical Association. She was a feminist who fought for equal opportunities and advocated sex education. The tablet was erected by the St Marylebone Health Society in 1933 at the place where Murrell in 1907 initiated infant consultation, which she continued for many years.

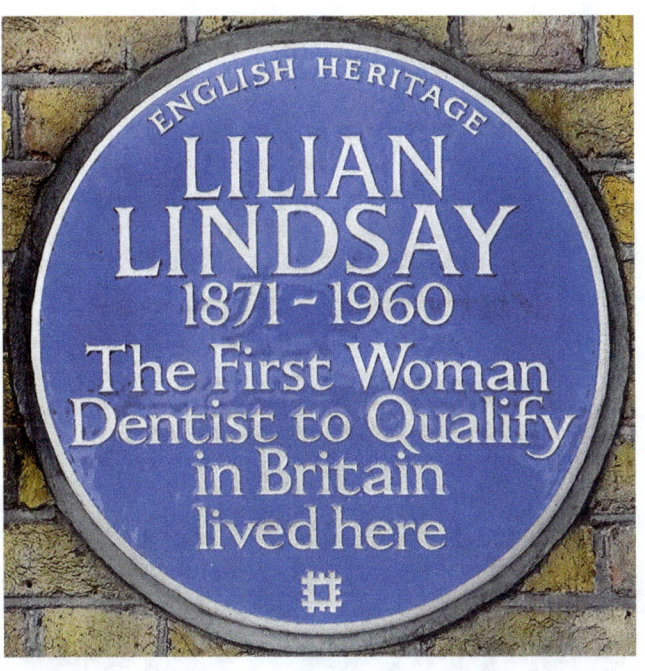

Lilian Lindsay (née Murray, 1871–1960) always wanted to become a dentist. She could not gain acceptance in London because she was a woman, so she moved to Edinburgh where she could study. Even there she sometimes experienced hostility as if she had taken the place of a man who would be the breadwinner for his family. Upon graduating from the Edinburgh Dental Hospital and School, in 1895, she became the first woman in the United Kingdom to qualify as a dentist. Later the same year, she was the first woman who joined the British Dental Association (BDA), then, half a century later, the first woman president of BDA. Other firsts preceded and followed this remarkable achievement. She was also active in establishing and enriching the Library of BDA and in initiating a museum of BDA. She published a book and many articles related to the history of dentistry in Britain. When, in 1920, her husband, Robert Lindsay, was appointed to a position in BDA, they moved to London and lived in an apartment above the BDA Headquarters on Russell Square. Her memorial plaque is now on the façade of this building.

Plaque of Lilian Lindsay, 23 Russell Square, WC1H (moved there from 3 Hungerford Road, N7, her former domicile; Spudgun67)

The first birth control clinics in Britain are commemorated by two plaques. Left: According to one, Dr. Marie Stopes opened the first clinic in 1921 in Holloway, and in 1925, it was moved to 108 Whitfield Street, W1T. Right: The other honors Dr. Charles Vickery Drysdale, founder of the Family Planning Association, which opened his first birth control clinic in 1921 at 153 East Street, SE17 (this image, Spudgun67)

Marie Stopes (1880–1958) was a paleo-botanist by training, who campaigned for eugenics and for women's rights. A dark side of her birth control activities was that they were primarily directed toward the poor and could be perceived as motivated by her dedication to eugenics. There is a memorial plaque (not shown here) on the house where she lived between 1880 and 1892, at 28 Cintra Park, SE19. It says that she was "Promoter of sex education and birth control." The plaque shown above commemorates the first birth control clinic opened by her. Today there is a network of clinics, "Marie Stopes UK."

Charles Vickery Drysdale (1874–1961), an electrical engineer and inventor, was co-founder of the Institute of Physics. However, he is best remembered for his activities as a social reformer. He opened his first birth control clinic in Britain in 1921, a mere 8 months following the one by Marie Stopes. It is remarkable that the two first birth control clinics opened almost at the same time.[4]

Plaques of Dame Ida Mann, 13 Minster Road, NW2, and Margery Blackie, 18 Thurloe Street, SW7 (both, Spudgun67)

Dame Ida Mann (1893–1983) was an ophthalmologist whose distinguished career started at a time when women seldom could embark on such a career. By 1927, she had a staff position at Moorfields Eye Hospital and a private practice in Harley Street. Her principal research concerned embryology and the genetic and social factors influencing the early development of the eye. She was involved with chemical defense during WWII. After the war, she and her husband moved to Australia, where she did much work for public health.

Margery Blackie (1898–1981) was a homeopathic physician who in 1969 was appointed physician to Queen Elizabeth II, the first woman doctor so distinguished.

---

[4]We thank Rocio Lale-Montes of the Institute of Reproductive and Developmental Biology, Imperial College, and Laura Sims of the Royal College of Obstetricians and Gynaecologists, for helpful exchange.

Left: Plaque of Dame Sheila Sherlock, 41 York Terrace East, Westminster NW1. Right: Painting of Dame Margaret Turner-Warwick by David Poole, 1992, Royal College of Physicians

Dame Sheila Sherlock (1918–2001) was declined repeatedly by several English medical schools in 1935–1936, until she was accepted by the University of Edinburgh. She graduated in 1941 at the top of her class. She attended the Royal Postgraduate Medical School, Hammersmith Hospital, in London (today, part of the Imperial College School of Medicine). Focusing her research on hepatitis, she became an internationally renowned authority of the liver. In 1951, she was the youngest woman to become a Fellow of the Royal College of Physicians. Then, in 1959, she was appointed to be Professor of Medicine at the Royal Free Hospital School of Medicine in London, the first ever female professor of medicine in the United Kingdom.

Dame Margaret Turner-Warwick (1924–2017) was a physician, most famous for her works on respiratory diseases. In her youth, she suffered from tuberculosis; thereafter, her interest settled upon lung function, cystic fibrosis, and asthma. Her recommendation of treatment with corticosteroids has been broadly followed. Among other contributions, she called attention to the dangers of working with asbestos.

Nurses

From left to right: Statue of Florence Nightingale, "The Lady of the Lamp," by Arthur George Walker (1915) as part of the Crimean War Memorial, Waterloo Place, SW1; her plaque, 10 South Street, W1K (Spudgun67); and her photograph by Kilburn (Wellcome Collection)

The nurse Florence Nightingale (1820–1910) is an iconic figure in British medical and military history. She revolutionized the treatment of the wounded in the Crimean War (1853–1856) and the functioning of hospitals. Four bas-relief panels decorate the stand of her statue at Waterloo Place depicting the following scenes: "Caring for the injured; Negotiating with military and political leaders; Challenging medical and hospital managers; and Teaching and inspiring nurses."

She assisted in the founding of the nursing school at St Thomas' Hospital, which was revolutionary in that it was a secular institution, and it was the first such nursing school worldwide. The graduates are known as Nightingales. There is a copy of her statue at the Central Hall of St Thomas' Hospital (see above). There is also a Florence Nightingale Museum on the campus of the Hospital at 2 Lambeth Palace Road, SE1.

From left to right: Mary Seacole's mosaic portrait on the wall of Morley College for adult education, 61 Westminster Bridge Road, Lambeth, SE1; her plaque, 14 Soho Square, W1; and her steel statue at the southern end of St Mary's Terrace, W2, as one of the local heroes

Here we display additional memorials to Mary Seacole whose principal memorial—a large statue—stands in front of St Thomas' Hospital (see above).

Left: Edith Cavell memorial by George Frampton, 1920, St Martin's Place, adjacent to Trafalgar Square. Right: Plaque of Ethel Gordon Fenwick, 20 Upper Wimpole Street, W1G

The nurse Edith Cavell (1865–1915) has memorials worldwide. She saved many lives in 1914–1915 in the war regardless of which side the wounded had been fighting for. She helped Entente soldiers escape from German-occupied areas, for which actions the Germans tried her for treason and executed her.

Ethel Gordon Fenwick (1857–1947) fought for improved recognition of nursing and the credential of Registered Nurse. When the Registration Act 1919 was accepted and the register opened in 1923, she was Nurse No. 1. She founded organizations in Britain and internationally, edited a journal of nursing, and established the British College of Nurses.

Professions

Anatomists

Of the three anatomists whose memorials we discuss in this chapter, those of William Harvey figure above.

Plaque of William Hunter, Great Windmill Street, Soho, W1V, and his statue by M. Noble at Burlington House, first from the left on the right side balustrade

William Hunter (1718–1783) was a leading anatomist and brilliant obstetrician. The surgeon John Hunter was his younger brother. William built his own anatomy theater and museum in Great Windmill Street where he trained the best anatomists and surgeons of his time. The Hunterian today is a world-famous museum, based on his rich collection. It houses, among many others, scientific instruments made by James Watt, Sir Joseph Lister, and Lord Kelvin. There is a modest rectangular memorial plaque in St James's Church, 197 Piccadilly Street, W1J, "To the memory of William Hunter, M.D., F.R.S., celebrated as a physician and physiologist."

Plaque of Henry Gray, 8 Wilton Street, SW1, and his portrait, engraved after a photograph by H. Pollock (Wellcome Collection).

Henry Gray (1827–1861) was elected Fellow of the Royal Society (FRS) at the age of 25. Devoted to anatomy, as a medical student, he had learned anatomy by dissecting cadavers rather than from books. He published his *Anatomy* in 1858 with the skilled artist Henry Carter's drawings. The book has been in print ever since. It is popular among medical students as well as among a wider audience. Gray died of smallpox although he had been vaccinated against it as a child with an early version of the vaccine. The American television series *Grey's Anatomy* is a far derivative of Gray's book, but this long-running dramatization of the lives of physicians, grappling with serious ethical issues, has inspired many young people to enter medicine.

Anesthetists

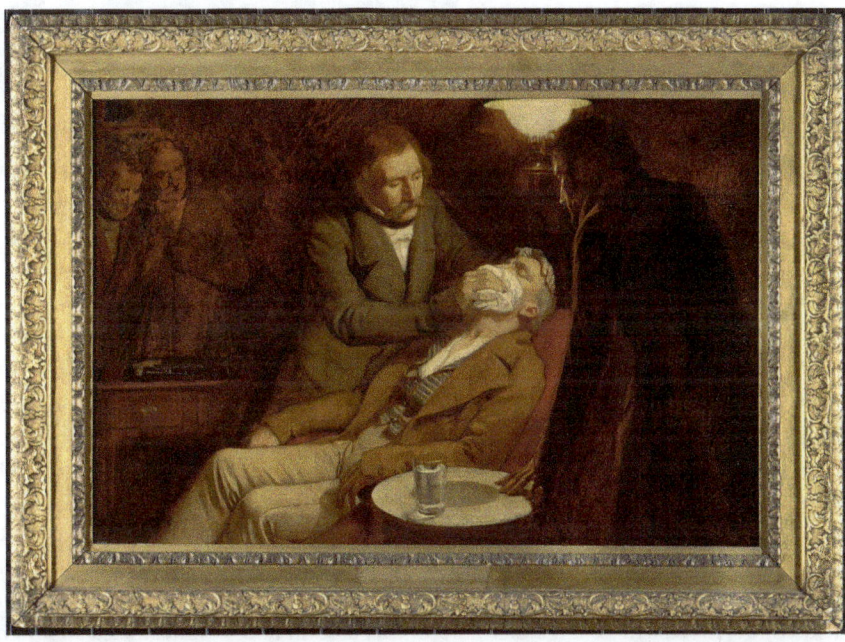

Portrait of William T. G. Morton and a painting showing the first use of ether in dental surgery, 1846, by Ernest Board (both, Wellcome Collection)

A landmark in anesthesiology, the first use of ether as an anesthetic, in dental surgery, is depicted in a painting by Ernest Board in the Wellcome Collection. Ether was administered by the American dentist William T. G. Morton (1819–1868), on September 30, 1846, in Boston, Massachusetts. Morton had tested the procedure on some animals and himself before he offered it to his patients. In order to entice a patient, Morton offered the patient $5 to let him try it.

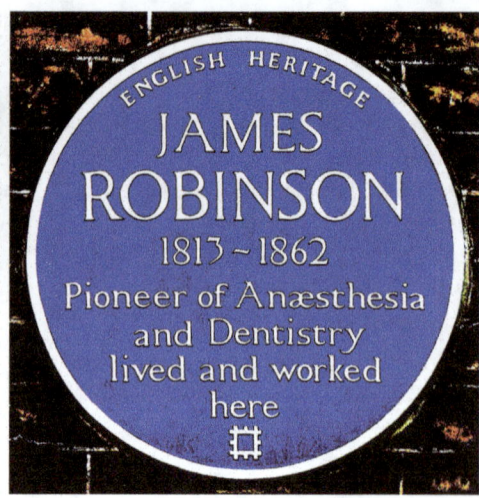

Left: Bonham Carter House, 52 Gower Street, WC1E with a blue plaque indicating the site of the first use of anesthetic. Right: Plaque of James Robinson, 14 Gower Street, WC1E

The news about Morton's innovation soon reached London. The first application of ether in Britain took place on December 19, 1846, in the Bonham Carter House. There, the dentist James Robinson (1813–1862) used it to ease a tooth extraction. Anesthesiology spread fast in the United Kingdom, and Robinson was recognized not only as the first but also as the best anesthesiologist in the country. He was so dedicated to this new approach that before 1847 was over he published a textbook about it. Soon, however, he returned to his original interest, dentistry, and encouraged his colleague John Snow to advance his studies in anesthesiology. Thus Snow also became a major figure in this area in addition to epidemiology for which he is primarily known (see below).

As was noted above (section of the Royal Society of Medicine), Sir James Young Simpson in Edinburgh pioneered the administration of ether during childbirth. Initially there was much resistance to the application of anesthesia at childbirth, for some religious fundamentalists argued that women should feel the pain based on scriptural precedent. But in 1853, Queen Victoria asked John Snow to administer chloroform when she was delivering her eighth child and 3 years later on another occasion. Her advocacy changed the general attitude and obstetrical anesthesia became accepted. It was also John Snow's contribution that administering the anesthesia should be done not by the surgeon, but by another person whose undivided attention would be directed to the procedure.

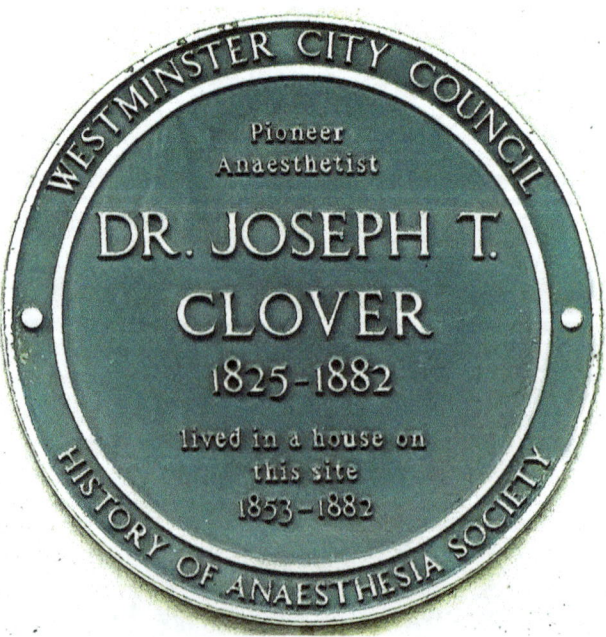

Portrait of Joseph Clover (Wellcome Collection) and his plaque, 3 Cavendish Place, NW2

Joseph Clover (1825–1882) was a physician who specialized in anesthesiology. He invented new instruments for improving the administration of the anesthetics during surgery. He designed bulky as well as portable apparatus and his portable version remained in use for decades. Clover administered anesthetics in the surgery of such famous people as Napoleon III; the future King Edward VII (then, the Prince of Wales) and Edward's wife, Princess Alexandra of Denmark; the politician Sir Robert Peel; and Florence Nightingale. Clover and John Snow are the two anesthetists who appear in the crest of the Royal College of Anesthetists.

The Anesthesia Heritage Centre provides ample information about the history of anesthesia. Its collection includes over 4,500 objects. The exhibition mounted in a small area is both rich in information and enjoyable, a good mix of personalities and instrumentation. We single out two pieces from the collection. One is the electrocardiogram (ECG) machine that was used in 1951 in a surgical operation upon King George VI (1895–1952). He was a heavy smoker and had lung cancer. When a malignant tumor was found in his lungs, the left lung was removed (pneumonectomy). The procedure took place in Buckingham Palace in a makeshift operating theater. The event was quite realistically recreated in the TV series *The Crown*, which used a team of real surgeons for the episode. The other exhibit is the Magill endotracheal apparatus. The anesthetist Sir Ivan Magill (1888–1986) introduced much innovation in modern anesthesia.

The ECG machine on display at the Anesthesia Heritage Centre, 21 Portland Place, Marylebone, W1B (courtesy of the Anesthesia Heritage Centre)

The Magill endotracheal apparatus on display at the Anesthesia Heritage Centre, 21 Portland Place, Marylebone, W1B (courtesy of the Anesthesia Heritage Centre)

Epidemiologists

Remains of the White Hart Dock as a Memorial of the Lambeth Cholera Epidemic along Albert Embankment

Before the Albert Embankment was built, it was the Lambeth's waterfront, an industrial hub with extremely high population density and poverty. By the early 1840s many took their drinking water from the terribly polluted River Thames at the White Hart Dock and similar locations. The river water was the main culprit of the cholera epidemic that killed 2000 people during a 1-year period starting in October 1848.

The pump memorial in front of the John Snow pub at the corner of Broadwick Street and Poland Street flanked by the view of the pub and its sign depicting John Snow

Plaques for John Snow on the building of the John Snow pub

Another serious cholera epidemic happened in 1854. A pump in Broad Street (today, Broadwick Street) in the Soho was identified as the notorious source of contaminated water. Now, this pump is no longer operational, but it stands there (in a slightly shifted location) without a handle. It is a solemn memorial to the many victims of the cholera epidemic and to the sagacity of the physician John Snow (1813–1858).

Central part of John Snow's map of Broad Street and vicinity in Soho and highlighting the area around the "Pump" (courtesy of Ralph R. Frerichs, UCLA Department of Epidemiology)

At the time cholera was considered an airborne disease, transmitted through a carrier called miasma. Snow drew maps, plotted the distribution of deaths, and zeroed in on the pump in Broad Street. Similar conclusions could be made about the Lambeth's waterfront. For his famous "detective work," John Snow is considered one of the founders of modern epidemiology. He did also pioneering work in anesthesiology. There are several memorials for Snow, but more recently, his role in fighting the cholera epidemic was dramatized in two popular venues: the Netflix series *Victoria* in which he appeared shy and with a stammer, but victorious, and the bestselling nonfiction book *The Ghost Map* (2006) by Steven Johnson.

Sir Edwin Chadwick (1800–1890) lived a long life, but his activities peaked during a relatively brief period 1832–1854, which overlapped with John Snow's. Chadwick was already involved with fighting the 1838 typhus epidemic and supported the notion that every household should have access to clean water. Similarly, he found it important to establish proper drainage and cleansing of the public sewers. He held offices with rather limited power for action, but his activities left their marks on progressive legislation concerning the protection of the poor and the improvement of sanitation in London. He authored legislation to have the dead removed from their homes and stored in a morgue until burial; in addition, he proposed that graves be 6 feet deep. His name is carved into the frieze of the London School of Hygiene & Tropical Medicine (see above).

Sir Joseph Lister Memorial, Portland Place, his portrait by Harry H. Salomon (Wellcome Collection), and a rectangular plaque at his former domicile, 12 Park Crescent

Sir Joseph Lister (1827–1912) pioneered antiseptic surgery. He attended UCL, which was then one of the few universities accepting Quakers, which was Lister's faith. He earned a Bachelor of Medicine degree and continued his studies at the Royal College of Surgeons. He started his career in Scotland. At the time, the general belief was that bad air caused the infections of wounds. Consequently, physicians, and surgeons among them, did not wash their hands before treating patients. In fact, a dirty surgical gown was considered a sign of long experience. This was so even though already years before, Oliver Wendell Holmes (Sr.) in the United States and Ignaz Semmelweis in Vienna showed that puerperal fever and the death of mothers were a consequence of unsanitary conditions in the delivery wards.

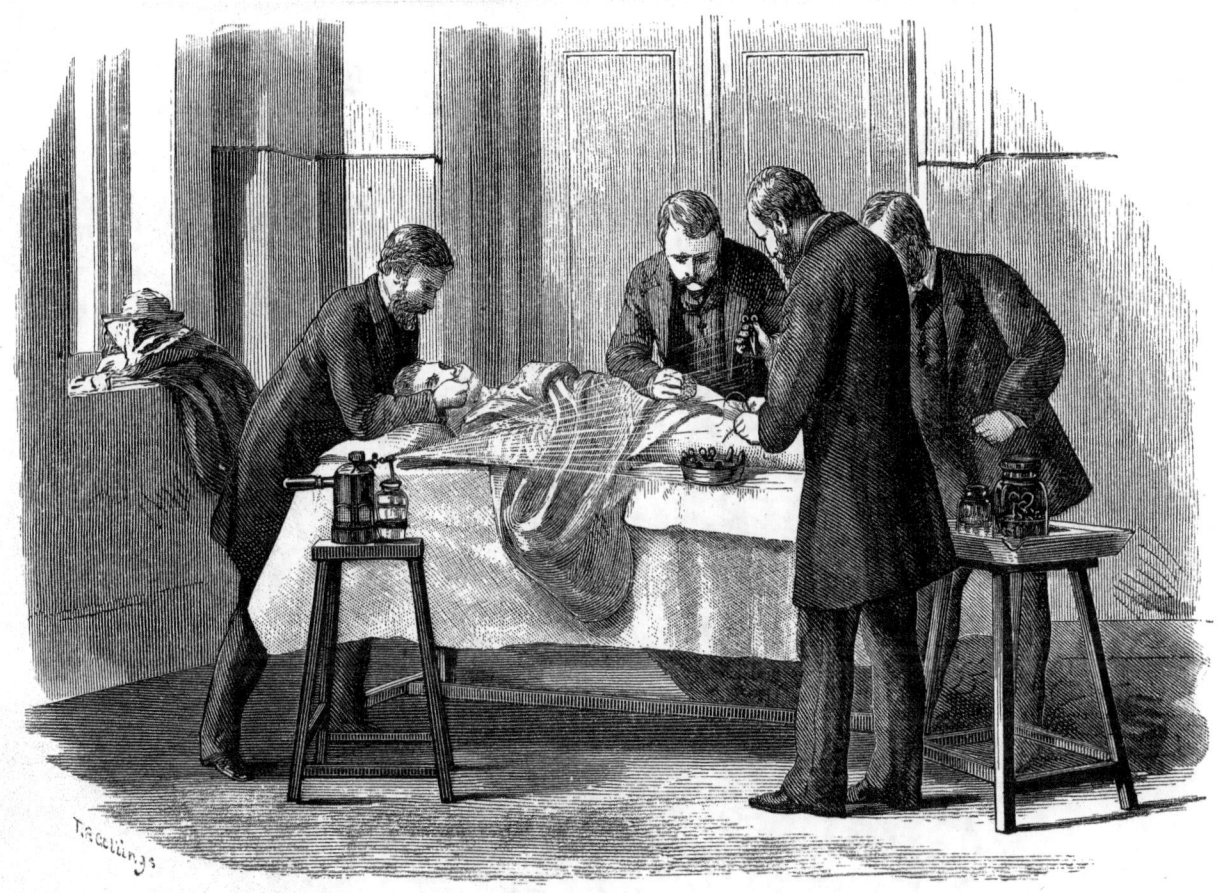

FIG. 23.

This figure represents the general arrangement of surgeon, assistants, towels, spray, &c., in an operation performed with complete aseptic precautions. The distance of the spray from the wound, the arrangement of the wet towels, the position of the trough containing the instruments, the position of the small dish with the lotion, the position of the house surgeon and dresser, so that the former always has his hands in the cloud of the spray, and the latter hands the instruments into the spray and various other points, are shown.

Use of the Lister carbolic spray—antiseptic surgery, 1882 (Wellcome Collection)

Lister was influenced by Louis Pasteur's teachings about the role of microorganisms in infections. Lister developed antiseptic techniques for treating wounds at the University of Glasgow. He published six papers about his experience in *The Lancet* in 1867. He prescribed for surgeons to wash their hands with 5% carbolic acid (phenol) solution before and after surgery, to spray their surgical instruments with such a solution, and to wear clean gloves. In spite of his demonstrable reduction in patient mortality, many in the medical profession did not accept his recommendations for years. Lister moved from Glasgow to Edinburgh, and then to London and worked for the rest of his medical career at the King's College Hospital. When his wife and long-time close associate in his work died in 1893, Lister ended his medical practice. However, when in 1902, Edward VII had to be operated on for appendicitis, the surgeons asked for Lister's advice and then followed his recommendations rigorously. The appendectomy was successful, and there was no postoperative infection. Named after Lister, the American mouthwash *Listerine®* was developed by a chemist, Joseph Lawrence, in 1879.

Immunologists

Portrait of Benjamin Jesty by Michael W. Sharp, 1805 (Wellcome Collection)

In the second half of the eighteenth century, a number of European physicians and laymen made observations about the deadly disease smallpox. In the course of epidemics, a large portion of those infected died. A few physicians noticed that prior exposure to an infection by cowpox, the milder animal disease, rendered the subject of infection immune to future smallpox infections. As many as six investigators in Britain, Denmark, and Germany purposefully administered infection with cowpox to protect those so inoculated from smallpox. Edward Jenner (see below) himself had so been inoculated in his childhood. The very first may have been the German Jobst Bose in Göttingen. Benjamin Jesty (c.1736–1816) may have been the second; he was certainly one of the pioneers. A farmer with a sharp eye and a dedicated experimenter, Jesty administered inoculations with the less virulent virus (of course, he had no notion about viruses; it is only that we can now apply such terminology). He lived in a village Yetminster, Dorset. He had already had cowpox by the time the next epidemic of smallpox arrived. However, he administered cowpox to his two sons and his wife by transferring pustular material from a sick cow to them, after abrading their arms. This happened in 1774. Jesty's neighbors ridiculed him for this experiment and they labeled it inhuman. However, Jesty was proved to be right and he saved his family from smallpox. He did not much publicize his deed; certainly he did not try to publish it, and it was only after Jenner's recognition that steps were taken to recognize Jesty. Part of this recognition was that he was invited to sit for a portrait, so posterity can be acquainted with him visually.

Statue of Edward Jenner by William Calder Marshall (1862) on the east side of the Italian gardens in Kensington Gardens (originally it was erected in Trafalgar Square, 1858) and a painting by Giulio Monteverde, showing Jenner vaccinating his son (Wellcome Collection)

Edward Jenner (1749–1823) was a physician and surgeon and the creator of the world's first vaccine, for smallpox. He grew up in Gloucestershire where between the ages 14 and 21 he apprenticed to a local surgeon. At 21, he continued at St George's Hospital in London where John Hunter was his mentor. He kept up his association with Hunter to the end of Hunter's life. In 1773, Jenner returned to Gloucestershire and took up his practice as a family doctor. He interacted with a broad circle of his colleagues, read the medical literature, and published papers on a variety of diseases and conditions, including smallpox. For an ornithological study of the nested cuckoo, he was elected Fellow of the Royal Society in 1788. He continued his studies and received the degree of MD in 1792 from what is today the University of St Andrews.

As Jenner was devoting himself to this topic, he observed that milkmaids showed remarkable immunity to smallpox. He hypothesized that the pus in the blisters that appeared as a consequence of infection served as their protection. In 1796, he used a sample of pus from such cowpox blisters on the hand of a milkmaid to inoculate an 8-year-old boy. The boy then became resistant to smallpox infection. The demonstration and reliable recording of this resistance and similar resistance showed by a number of others Jenner had inoculated was the essence of Jenner's discovery. Once he proved the validity of his observation, he reported them to the Royal Society and published them. This publication of his discovery has given Jenner priority over Jesty. In time, Jenner's discovery was accepted, and, eventually after enacting the necessary legislation, widespread vaccination was introduced. Even though this recognition came years after Jenner's death, he still enjoyed the fame and international recognition that came in his lifetime. Jenner's activities laid the foundation of the modern science of immunology. Although there have been numerous memorials erected in his honor, his truest recognition is in the tens of millions of lives the vaccination protocols initiated by him saved.

Plaque of Sydney Monckton Copeman, 57 Redcliffe Gardens, SW10 (Spudgun67), and his portrait (Wellcome Collection)

Regarding the smallpox vaccination, the contribution of Sydney Monckton Copeman (1862–1947) must be mentioned. He studied at Cambridge and St Thomas' Hospital and went into research at the encouragement and financial backing by Sir Joseph Lister. Copeman worked for the Local Government Board and the Ministry of Health and produced an improved version of the smallpox vaccine. Due in large part to his efforts, in 1979, the World Health Organization declared smallpox to have been eradicated.

 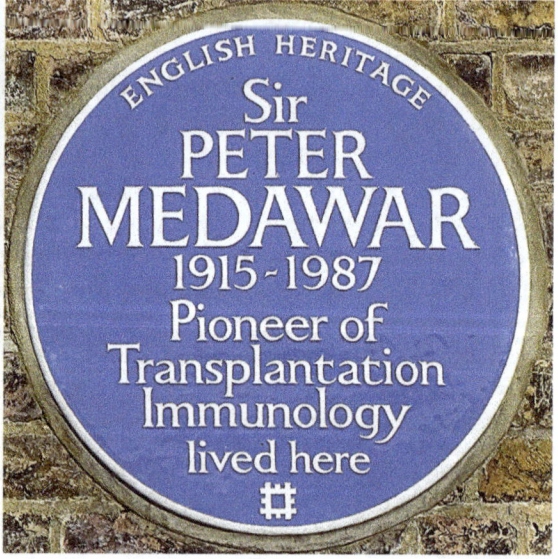

Plaque at Sir Peter Medawar's Tree at the north-west entrance into the garden of Soho Square, W1, and his plaque, 25 Downshire Hill, Hampstead, NW3

Sir Peter Medawar (1915–1987) studied at Magdalen College in Oxford, majoring in zoology. He had a broad interest, but with World War II his ambitions in science made a sharp turn in that he dedicated himself to helping burn victims, among them RAF wounded. He investigated the immunological reactions to skin transplants, which was similar to the reaction of the organism to foreign invaders. In 1960, he was awarded the Nobel Prize in Physiology or Medicine jointly with the Australian Macfarlane Burnet for their independently made discovery of acquired immunological tolerance.

Laryngologists

Left: Plaque of James Yearsley, 32 Sackville Street, W1, where he founded the Metropolitan Ear Institute in 1838. Right: Plaque of Sir Morell Mackenzie, 33 Golden Square, W1, where he founded the world's first hospital for diseases of the throat in 1865

James Yearsley (1805–1869) studied medicine at St Bartholomew's Hospital after having apprenticed with a surgeon. Eventually, he became a member of the Royal College of Surgeons and other learned institutions of eminence. He trained himself to be an aural surgeon and in 1846 took the position of surgeon to the Royal Society of Musicians. He greatly lifted the level of scholarship in aural surgery. He recognized the connection between deafness and the pathological condition of nasopharynx, the pharynx being part of the digestive and the respiratory systems. He devoted much effort to find ways to diminish deafness.

Sir Morell Mackenzie (1837–1892) studied in Walthamstow, King's College London, and the London Hospital. Qualifying for membership of the Royal College of Surgeons in 1858, he continued his education in Paris, Vienna, and Budapest and became acquainted with the new laryngoscope. From 1862, he was at the London Hospital, and in 1863, he co-founded the Hospital for Diseases of the Throat. He became internationally recognized in this area of medicine and was asked for consultation when the Crown Prince, later, Emperor Friedrich III, of Germany suffered from laryngeal cancer.

Plaques of Sir John Milsom Rees, 18 Upper Wimpole Street, W1 (left), and Lionel Logue, 146 Harley Street, SW1 (right)

Sir John Milsom Rees (1866–1952) was a surgeon specializing in laryngology. He studied medicine at St Bartholomew's Hospital. Following stints in Edinburgh and Tottenham, he set up private practice in London and was appointed consultant to the Royal Opera House, Covent Garden. Famous singers were among his patients. In 1910, he became the laryngologist of the Windsors and treated George V and other members of the royal family.

The Duke of York, later, King George VI, had a problem in the coordination between his larynx and his thoracic diaphragm, resulting in speech impediment that appeared as a stammer. Encouraged by his wife, the Duke turned to the speech therapist Lionel Logue (1880–1953) for assistance. Logue grew up in Australia, where he learned, then taught, elocution. In 1924, he and his family moved to London and he opened his practice of speech therapy at 146 Harley Street. This is where he treated the Duke of York. When the Duke became King, their sessions continued and Logue helped the Sovereign through the years of World War II, when there was so much need for his public speaking, especially radio addresses. Their relationship is beautifully depicted in the film *The King's Speech*.

Ophthalmologists

Sculpture relief by Eric Gill over an entrance to the 1933 extension of the Moorfields Eye Hospital, 162 City Road, Old Street, EC1V

The sculpture relief decorating the façade of the Moorfields Eye Hospital depicts a biblical scene. Bartimaeus was a blind man whom Jesus Christ healed so he could see (Mark 10:46–52). When Jesus asks Bartimaeus what he should do for him, the blind man's response is "*Domine ut videam*" ("Lord, make me see")—words carved into the memorial.

Plaques of Sir William Bowman at St Antony's School, Ivy House, 94–96 North End Road, Golders Green, NW11 (Spudgun67), and Sir Stewart Duke-Elder, 63 Harley Street, W1

Sir William Bowman (1816–1892) was an ophthalmologist who excelled as a surgeon and an anatomist as well. He studied at King's College in London and assisted the physiology professor, Robert Bentley Todd, in his dissections for demonstration. Later, Bowman and Todd co-authored several volumes in physiology. Bowman made discoveries in the anatomy of the kidney, the olfactory mechanism, and the eye. He worked at the hospital, later known as the Moorfields Eye Hospital, and founded a society, later known as the Royal College of Ophthalmologists. A memorial tablet honors him within St James's Church, Piccadilly.

The ophthalmologist Sir Stewart Duke-Elder (1898–1978) researched the impact of intraocular pressure in the eye. In this he was a forerunner of the discoveries that led to a successful treatment of glaucoma toward the last decades of the twentieth century. He was a prolific author and editor and served as the surgeon-oculist to kings and to Queen Elizabeth II.

Left: Plaque of Josef Dallos, 18 Cavendish Square, W1. Right: George Nissel plaque, at the corner of Siddons Lane and Melcombe Street, NW1 (this image, Matt Brown)

Josef Dallos (1905–1979) became an ophthalmologist under the internationally renowned Professor Emil Grósz (1865–1941) at the first Clinic of Ophthalmology of the University of Budapest. Shortly after the Jewish Grósz's forced retirement in 1936, Dallos immigrated to the United Kingdom. Already in Budapest he had become a recognized specialist on contact lenses and Ida Mann (more about her above, in Women in Medicine) was among his international visitors. In London, he helped servicemen with scleral contact lenses in WWII and those who suffered from WWI gas poisoning of the eye. The scleral lenses are large contact lenses made of blown glass that are adjusted to the sclera (the white outer layer of the eyeball). As an additional benefit, they help keep the eye moist. The memorial plaque recognizes his pioneering work with contact lenses.

The Transylvania-born George Nissel (1913–1982) studied engineering in Brno, Czechoslovakia.[5] He married Dallos's sister and the Hungarian-Jewish family immigrated to the United Kingdom in 1937. When they were yet in Budapest, Nissel became fascinated with contact lenses that Dallos's skillful technician, Istvan Rakos, was preparing for the ophthalmologist. Nissel started making scleral lenses for Dallos in London during WWII. After the war, he extended his business and supplied his lenses all over in the United Kingdom and exported them to Sweden. As his activities broadened, he had special lathes and polishing machines prepared for his optical work. He issued a newsletter periodically and was an active member of the contact lens community worldwide.

---

[5] Between the two world wars, hundreds of Hungarian Jews attended the German Technical University in Brno, Czechoslovakia (as it was then), which is today part of Masaryk University. They were prevented from receiving higher education in Hungary due to the infamous 1920 Law, known as *numerus clausus*, severely restricting the number of Jewish students in Hungarian higher education.

Pathologists

Plaques of Sir Bernard Spilsbury, 31 Marlborough Hill, NW8, and Cedric Keith Simpson, 11 Weymouth Street, W1 (both, Spudgun67)

The pathologists, Sir Bernard Spilsbury (1877–1947) and Cedric Keith Simpson (1907–1985), were also famous forensic medical experts of their time. A third pathologist, Thomas Hodgkin, will be discussed below. Spilsbury was involved in many trials that caused great excitement among the general public. He devised the "murder bag," which was a forensic kit for detectives investigating unnatural death cases used to preserve the integrity of evidence. He was instrumental in the British deception "Operation Mincemeat" in World War II. A cadaver dressed as a British officer, carrying impressive documents, was planted in the Mediterranean Sea so as to mislead the Germans about the location of the Allied invasion in Southern Europe. Simpson was a professor of forensic pathology and a prolific author. He taught at Guy's Hospital, the University of London, and the University of Oxford. After the war, he did pioneering work for recognizing battered baby syndrome.

Pharmacologists

Plaque of Sir Henry Dale on the stone fence of Mount Vernon House, Mount Vernon, NW3, and his portrait (Wellcome Collection)

Sir Henry Dale (1875–1968) was a physiologist and pharmacologist. In 1914, he showed that the substance called acetylcholine was the means of chemical signal transmission of nerve impulses. The 1936 Nobel Prize in Physiology or Medicine was shared between Dale and Otto Loewi, who demonstrated the importance of Dale's discovery for the understanding of the nervous system. Dale had a spectacular career in British scientific life, with positions which included, among others, President of the Royal Society of Medicine, Chairman of the Wellcome Trust, and President of the Royal Society.

Dale recognized early the significance of the discovery of Oswald Avery and his two young associates in 1944 that DNA was the substance of heredity. Avery did not receive the Nobel Prize, and only years after his death did the Nobel Foundation, in a rare admission of error, concede that Avery had deserved the award. In contrast, Dale, as President of the Royal Society, distinguished Avery with the highest recognition of the Society, the Copley Medal, in 1945. Dale stated, "Here surely is a change to which, if we are dealing with higher organisms, we should accord the status of a genetic variation; and the substance inducing it—the gene in solution, one is tempted to call it—appears to be a nucleic acid of the desoxyribose type. Whatever it be, it is something which should be capable of complete description in terms of structural chemistry."[6] This was foresight. Enunciating it required courage on Dale's part.

---

[6] H. Dale, Anniversary Address to the Royal Society, 1945, pp. 1–17; the actual quote is on p. 2.

Left: Plaque on the façade of the Alexander Fleming Laboratory where Fleming made his milestone discovery. Right: Sign of the Alexander Fleming Laboratory Museum at the St Mary's Campus of Imperial College (corner of Praed and Norfolk Streets)

Left: Memorial plaque with a Fleming bas-relief by Tim Metcalfe (1993) at the St Mary's Hospital Medical School, corner of South Wharf Road and Norfolk Place, W2. Right: "Fleming in his laboratory" stained glass window by Arthur Buss (1952) in St James's Church, Sussex Gardens, Paddington, W2

Left: The Fountains Abbey pub, 109 Praed Street, Paddington, W2. Right: Fleming poster displayed in the pub.

Sir Alexander Fleming (1881–1955) was a bacteriologist whose initial discovery of penicillin has been retold many times as a romanticized story. His fate and road to the Nobel-Prize-winning discovery is extraordinary, with some moments in it that are most instructive. He was 13 when his father died and the family was quite impoverished. He was sent from Scotland to London to learn some trade, where he enrolled in the Polytechnic Institute and then worked as a clerk in a shipping office. Obviously, he was aiming at something else. He applied for and received a scholarship, and in 1902 he enrolled for medical studies at the St Mary's Hospital Medical School from which he graduated in 1906. He joined St Mary's bacteriology department and stayed there for his entire career, except for the period 1914–1918. During World War I, he served as captain in the British Army Medical Corps.

He was in the forefront of research and was the first who injected anti-typhoid vaccines into humans. He was also among the first physicians who used *Salvarsan* against syphilis, which had been discovered by Paul Ehrlich. Fleming's ambitions showed when he embarked on finding the pathogen of the common cold. He did not succeed; instead, he discovered the enzyme lysozyme in tears and mucus. This was the first antibiotic he discovered, but the bacteria that lysozyme destroyed represented no harm to humans. This was in the early 1920s. In 1928, he was working with some cultures of the bacterium *Staphylococcus aureus* when one day he noticed that in one of the containers—a Petri dish— green mold was thriving around where the bacterial colonies had disappeared. As he was investigating the bacterial colonies, he might have been upset by the molds for having destroyed his sample. A less observant experimenter might have thrown out the "infected" Petri dish. Instead, he wondered what the substance that killed his bacteria might be. He performed some experiments and noted that the mold was similar to the common one that grows on stale bread. He coined the name penicillin for the active antibiotic component, and the strain was later identified as *Penicillium notatum*.

His observations were published in 1929. He did not embellish the story by concocting a sophisticated narrative reflecting deep thoughts and careful planning. Indeed, Fleming made an observation by chance, and his main merit was that rather than dismissing the result, he published it. Further work was needed to purify penicillin and produce it in sufficient quantities for it to become a medicine. The pharmacologist and pathologist Howard Walter Florey (1898–1968), the chemist Ernst Boris Chain (1906–1979; more about him in the Science chapter), and the biochemist Norman Heatley (1911–2004) took up this task in 1939 at Oxford University. They solved the processes of growing and purifying penicillin. There was a race for satisfying the rapidly growing demands for penicillin, and the raging world war further multiplied those demands. In 1945, Fleming, Florey, and Chain jointly received the Nobel Prize in Physiology or Medicine for penicillin. There was no possibility for a fourth Nobel award. Heatley's main recognition came belatedly, in 1990, when Oxford University recognized him with an honorary degree of Doctor of Medicine. This was the only time that Oxford awarded a non-medic a medical doctorate.

Fleming has become something of a folk hero. There is a Fleming cult in the Fountains Abbey pub across the street from his former laboratory. A big poster quotes Fleming as saying "One sometimes finds what one is not looking for." Within his lifetime, Fleming received the knighthood, 25 honorary doctorates, 26 medals, 18 prizes, 13 decorations, and honorary memberships in 89 scientific societies.[7] The others were less prone to media connections. But the most substantial legacy of the four discoverers was the millions of lives saved by penicillin.

---

[7] Paul Brack, "Norman Heatley: the forgotten man of penicillin." *Biochemistry*, October 2015, pp. 36–37.

Physicians

Statue of "Healing" or "Care of the Sick" by Alfred Hardiman, 1932, and the central section of the former London County Hall. The statue is on the façade of the river front, northern end pavilion

"Healing" or "Care of the Sick" are titles that well express the calling of physicians. The statue shown here is one of ten monumental sculptures decorating the former London County Hall. Each of them was supposed to stand in the section of the building in which the county officials were dealing with the tasks appositely expressed by the respective statue.

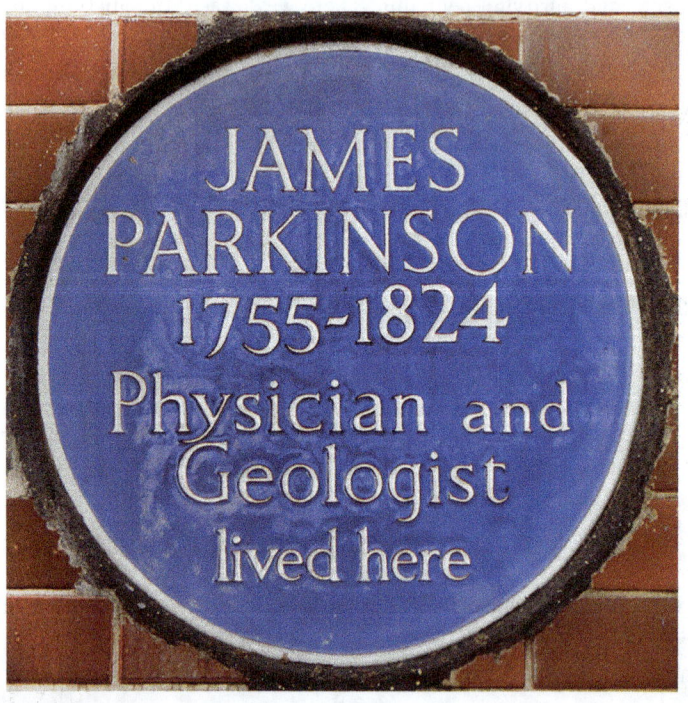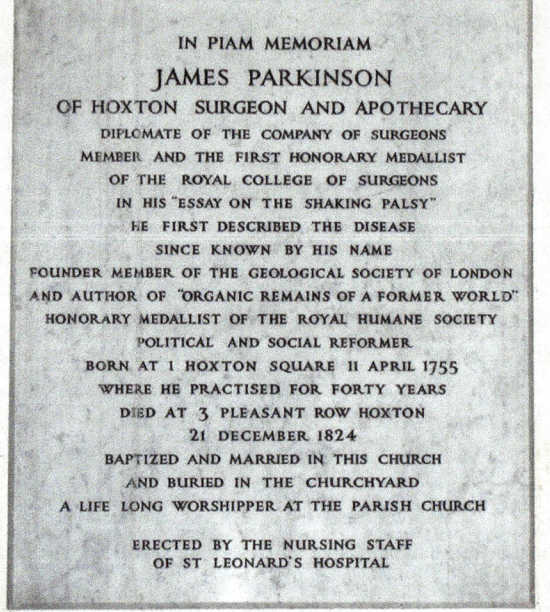

James Parkinson's memorial plaque, 1 Hoxton Square, N1, and tablet in St Leonard's Church in Hoxton (both, courtesy of Patrick Lewis)

James Parkinson (1755–1824) enrolled at the London Hospital in 1776 and graduated in 1784 at the Company of Surgeons, the predecessor of the Royal College of Surgeons. He worked as a general practitioner in his birthplace, Hoxton, and was also interested in the patients of the mental health institutions in and around Hoxton. His most important contribution to medicine was in neurology. He described the condition of shaking palsy, which today is known as Parkinson's disease. Beside medicine he published significant treatises in paleontology.

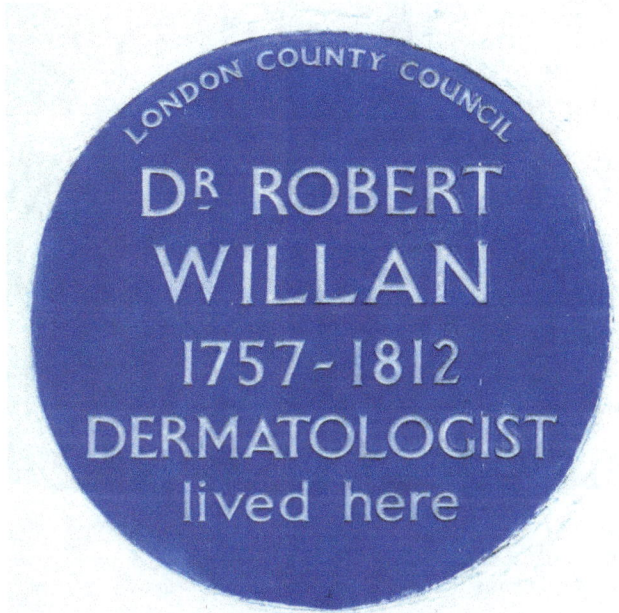 

Plaques of Robert Willan, 9–11 Bloomsbury Square, WC1, and Edward Meryon, 17 Clarges Street, W1

Robert Willan (1757–1812) earned his qualifications in medicine in Edinburgh, moved to London in 1783, and became a dermatologist. He classified the diseases of the skin and has been considered as the founder of dermatology. Another of his pioneering inquiries was eating disorders. He investigated the occupational disease *psoriasis diffusa* affecting bakers. He was constantly on the lookout for uncharted domains in medicine.

Edward Meryon (1807–1880) described as early as 1851 the pathological condition usually referred to as the Duchenne muscular dystrophy. It is a genetic disorder of progressive wasting and weakening of the muscles from early childhood and leading to early death. The disease is ascribed to the deficiency of the membrane protein dystrophin. The French neurologist Guillaume-Benjamin-Amand Duchenne de Boulogne (1806–1875) described it in the 1860s, but Meryon's priority has been documented meticulously.

Left: Statue of Robert Bentley Todd by Matthew Noble, 1862, in front of King's College Hospital, Bessemer Road, Denmark Hill, SE5. Right: Medallion depicting John Langdon Down, 1997, on the façade of the former Normansfield Training Institution for Imbeciles at the corner of Kingston Road and Normansfield Avenue, TW11. Photograph by and courtesy of Ian Jones-Healey, the Langdon Down Museum

Robert Bentley Todd (1809–1860) was first licensed at the Royal College of Surgeons in Ireland in 1831, then, at the Royal College of Physicians in London in 1833. From 1836, he was Professor at King's College. His main interest was in physiological medicine—at the time a fledgling area of health science. He described the postictal state, which occurs when following an epileptic seizure, the patient experiences an altered state of consciousness. Sometimes it is called Todd's palsy.

John Langdon Down (1828–1896) started as his village apothecary father's apprentice at the age of 14, in Northern Ireland. At 18, he was already in London working as a surgeon. He continued in a pharmaceutical laboratory where he excelled in organic chemistry. At some point he helped Michael Faraday with experiments on gases. When his father recalled him, he continued at the apothecary until his father passed away in 1853. Only then, Langdon Down's career began. He studied at the Royal London Hospital and qualified in 1856 at the Apothecaries Hall and the Royal College of Surgeons. He worked at the London Hospital and at what is today the Royal Eastwood Hospital, but was then an "asylum for idiots and institution for the mental defectives." He and his wife greatly improved the conditions at Eastwood. He published papers about his observations of his patients. In particular, he described and classified what was then called "the mongoloid type of idiot," but is today called Down syndrome after Langdon Down. Although today the earlier term sounds racist and hurtful, he took a stand for the unity of the human species and against regarding different races as separate species. He set up a private practice at Normansfield and today there is the Langdon Down Centre.

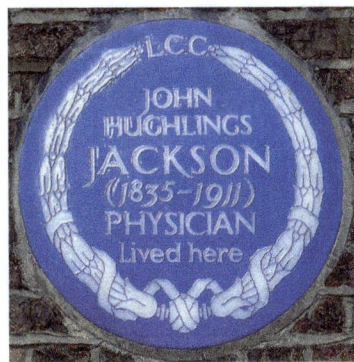   

Extreme left and middle left: John Hughlings Jackson's plaque, 3 Manchester Square, Marylebone, W1U, and bust at the Institute of Neurology, Queen Square, WC1 (Wikipedia). Middle right and extreme right: Sir James Mackenzie's plaque, 17 Bentinck Street, W1, and portrait (Wellcome Collection)

John Hughlings Jackson (1835–1911), a neurologist, studied in York and London and worked mostly at what is now the National Hospital for Neurology and Neurosurgery, where he followed broad research interests and was a prolific author. His principal contributions were to the understanding and treatment of epilepsy. He was a co-founder of the journal *Brain*.

Sir James Mackenzie (1853–1925) was a cardiologist. He studied in Edinburgh, spent some time in London, initiated the *British Heart Journal*, and founded the Mackenzie Institute of Clinical Research in St Andrew's to the benefit of not only specialists but also of primary caregivers. He pioneered research into cardiac arrhythmia and he himself suffered from the condition. Before he died, he asked his friend, John Parkinson, to perform his autopsy. Parkinson complied, but the heart needed further investigation by a specialist in Mackenzie's Institute. The findings were reported in 1939 in the *British Heart Journal*. The author, David Waterston, made an interesting comment in the paper, noting that Mackenzie, "like so many other doctors, had not been under the care and observation of a medical man from the commencement of his illness."[8]

"Mother and Child" by Patricia Finch, 2001, Queen Square, WC1. It is dedicated to the memory of Andrew Meller by "The Friends of the Children of Great Ormond Street Hospital."

---

[8] David Waterston, "Sir James Mackenzie's heart." *British Heart Journal* 1939, 237–248; the quotation is from page 237.

Plaques of Sir George Frederic Still, 28 Queen Anne Street, W1, and Grantly Dick-Read, 25 Harley Street, W1

Sir George Frederic Still (1868–1941) was from a working-class background and could acquire higher education only thanks to scholarships. He studied in Cambridge and at King's College London School of Medicine. He was the first professor of childhood medicine in England at the Great Ormond Street Hospital for Children. He described a form of childhood febrile arthritis (Still's disease) and a functional heart murmur in young children (Still's murmur). He was the first who recorded the symptoms of what is known now as the attention-deficit/hyperactivity disorder or ADHD.

Grantly Dick-Read (1889–1959) was an obstetrician who considered his lifelong mission to promote natural childbirth. He summarized his message in his last book, *Childbirth Without Fear*. He nurtured some old-fashioned views about the women's principal role in society when he wrote, for example, that the mother is a factory whose efficiency as such can be enhanced by proper education.

Psychiatrists and Psychologists, Including Psychoanalysts

 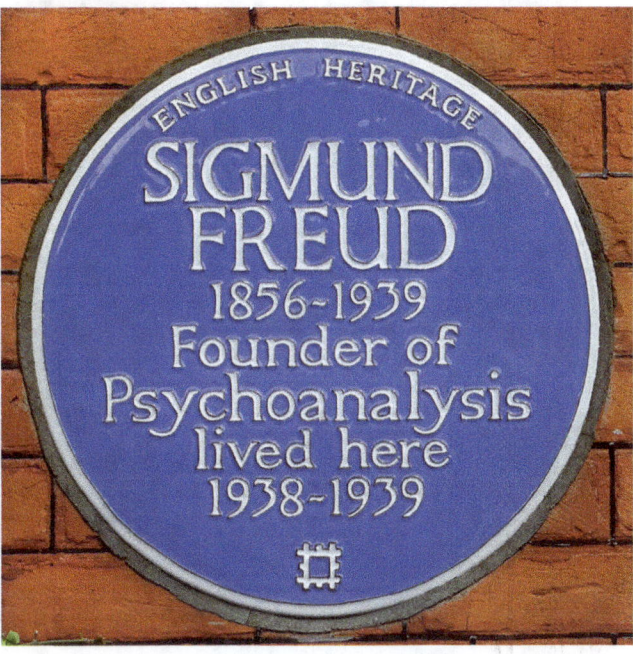

Statue of Sigmund Freud by Oscar Nemon, 1971, at the junction of Fitzjohn Avenue and Belsize Lane, NW3, in front of Tavistock Clinic, and his plaque at 20 Maresfield Gardens, NW3 (this image, Spudgun67)

Sigmund Freud (1856–1939) was an Austrian neurologist who founded psychoanalysis. He studied in Vienna and became a doctor of medicine. To this date he has remained influential and much cited. Freud followed a seemingly simple approach of treating his patients of psychopathology—having conversations with them. The noted Austrian sculptor, Oscar Nemon, was working on his Freud statue when he had to stop it as both Nemon and Freud had to flee the Nazis in 1938 upon the annexation of Austria by Germany (the *Anschluss*). The 82-year-old Freud found his new home in London, but did not live long. Nemon's Freud statue was erected in London. There is a memorial plaque on Freud's last home, which today houses the Freud Museum. Eighty years after Freud was forced out of Austria, in 2018, a copy of the London Freud statue was unveiled at the Medical University of Vienna, a belated expression of honor by the Viennese toward one of its greats.

Plaque of Ernest Jones, York Terrace East, NW1, and a group of psychoanalysts in 1909; from left to right, back row: Abraham A. Brill, Ernest Jones, and Sándor Ferenczi; front row: Sigmund Freud, G. Stanley Hall, and Carl Jung (Wellcome Collection)

Ernest Jones (1879–1958), neurologist, was the first English-speaking psychoanalyst. He studied at Cardiff University in Wales and at UCL. Eventually, he earned membership in the Royal College of Physicians. During the 1900s, Jones met Freud's disciples and Freud, and a friendship and professional interaction developed between them even if it was marred years later by controversy. Jones's initial attempts of doing psychoanalysis were made in North America. Then, in 1919 he founded the British Psychoanalytical Society and was active in setting up other institutions and publications to promote the new field. Following the Nazi takeover of Germany, Jones helped find positions for dismissed Jewish scientists. After the *Anschluss*, Jones flew to Vienna to help Freud and his family to immigrate to London. At Freud's funeral, Jones and the Austrian writer Stefan Zweig gave the eulogies. Jones acted as Freud's official biographer and his monumental Freud monograph of three volumes appeared after Jones's death, in 1959.

Left: Plaque of Melanie Klein, 42 Clifton Hill, NW8 (Spudgun67). Right: Plaque of Anna Freud, 20 Maresfield Gardens, NW3 (Spudgun67)

Melanie Klein (née Reizes, 1882–1960) and Anna Freud (1895–1982), Sigmund Freud's daughter, were Austrian-born British contributors to and pioneers of child psychoanalysis.

 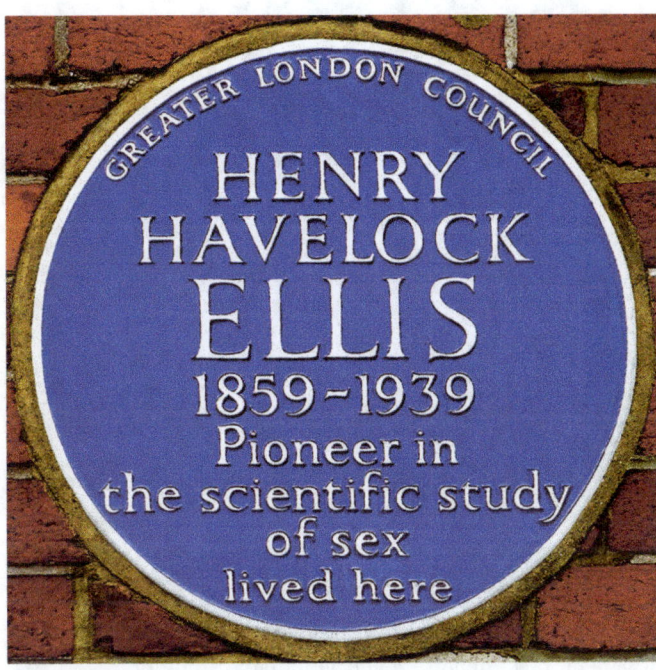

Henry Havelock Ellis, 1914 (Wellcome Collection) and his plaque, 14 Dover Mansions, SW9 (Spudgun67)

Henry Havelock Ellis (1859–1939) studied medicine at St Thomas' Hospital. From the beginning he intended to research sex and graduated as a physician to have the necessary credentials for his further studies. He pioneered investigations of homosexuality and transgender psychology, introduced such notions as narcissistic behavior and autoeroticism, and co-authored the first English medical book on homosexuality, *Sexual Inversion* (1896). He did not consider homosexuality a disease or a crime, as was the prevalent view at the time.[9] He initiated studies of psychedelic drugs that cause changes in the visual and auditory sensations and in consciousness. He supported eugenics and for some time was active in the movement until he fell out with the organization as he opposed sterilization.

---

[9] Writer Karl-Maria Kertbeny had coined the terms *heterosexual* and *homosexual*, only a few years before.

Plaque of Michael Balint and Enid Balint, 7 Park Square West, NW1

The couple Michael Balint (1896–1970) and Enid Balint (1903–1994) worked as psychoanalysts. They, together with a group of general practitioners, conducted a large study of doctor-patient relationship in the 1950s. Their recommendations influenced beneficially the process of diagnosis and treatment of patients by general practitioners. To mark the 50th anniversary of the formation of the Balint Society in 1969, and honoring the Balints' work, a memorial plaque was unveiled on the façade of the building where they lived and worked.

Michael Balint was a graduate of what is today the Semmelweis University in Budapest and studied psychoanalysis with Sándor Ferenczi. The autocratic and anti-Semitic Horthy regime in Hungary did not tolerate Balint's science, so he fled to the United Kingdom in 1938. First, he practiced in Manchester, then, he moved to the Tavistock Institute in London. There he met his future wife, Enid, and they jointly developed their most fruitful research projects and practices. At the time of Michael's death, he was the President of the British Psychoanalytical Society.

Surgeons

Statue of Galen (Claudius Galenus) by J. S. Westmacott on the central balustrade of Burlington House

Aelius (or Claudius) Galenus ("Galen," as inscribed in the anglicized way at Burlington House, c.130–c.201) was a Greek physician in the Roman Empire. He was a physician to gladiators and later to emperors. He was a prolific author who also wrote on philosophical and medical topics. What we know about medicine of his time is from his writings. He based the treatment of his patients on observation and experimentation; he dissected animals and treated wounded gladiators. He gathered a great deal of knowledge thereby about the physiology of the heart, the liver, and the brain. He was the first to observe the pulse for diagnostic purposes. Extensive literature exists about his life and works and about his impact on the further development of medicine, which lasted for centuries.

Memorials of John Hunter, from left to right: bust by N. F. Boonham in Lincoln's Inn Fields, WC2; plaque in the basement of St Martin's Church; and plaque on the façade, 30 Golden Square, Soho, W1F

John Hunter (1728–1793) was a surgeon, the founder of scientific surgery. He was the younger brother of William Hunter (see above). John learned a great deal of medicine from William, whom he assisted in his anatomic dissections. John himself built up a large anatomic collection of animal skeletons. He investigated blood in connection with the frequent bloodletting practice employed for a variety of diseases. He concluded that inflammation was a response of the organism to disease rather than being a disease itself—a view that is still an important tenet in modern medicine. John Hunter studied at Chelsea Hospital and the St Bartholomew's Hospital and worked as a surgeon at St George's Hospital. For a few years he served as an army surgeon. After he left the service, he established his private practice. He was considered a leading expert on venereal diseases. In 1776, he was appointed surgeon to George III. Later in his career, he became surgeon general and reformed the system of appointments of army surgeons to be based on merit rather than patronage. John Hunter was one of Edward Jenner's mentors, and later they were colleagues. His remains were in St Martin's Church from 1793 to 1859 when after proper identification they were transferred to Westminster Abbey. Apart from his bust at the Hunter Museum (named in his honor) of the Royal College of Surgeons, there is a John Hunter bust at the main entrance to the St George's Museum in Tooting, South London, and there used to be yet another bust on Leicester Square.

Plaques of Thomas Wakley and Thomas Hodgkin, both at 35 Bedford Square, Bloomsbury, WC1B

Thomas Wakley (1795–1862) was a surgeon and the founder of *The Lancet*. He started the weekly medical journal in 1823, and, through his sons, it stayed within the family for two generations after his death. It was an instant success and had already a circulation of 4000 by 1830. It is still one of the most prestigious science publications. *The Lancet* has remained independent and has greatly expanded over the years. It is no longer a single journal; rather, it is a family of medical periodicals, a great success story.

Thomas Hodgkin (1798–1866) began his way to the medical profession by studying apothecary. At the age of 21, he enrolled at St Thomas's and Guy's Medical School. He then continued at the University of Edinburgh where he received his MD degree in 1823. From 1825, he worked at Guy's Hospital in London where he was engaged in what today is called anatomical pathology. He was a great advocate of the importance of postmortem examination. He left Guy's in 1837 and established private practice. He interacted with Sir Joseph Lister in research. Hodgkin is best known for the disease he described in 1832 and has been called Hodgkin's lymphoma from around the time of his death. It is a malignant inflammation of the lymph glands, often with simultaneous damage to the spleen and the liver. It usually occurs in young adults. His name also figures in the so-called non-Hodgkin lymphoma, a blood cancer, which includes all types of lymphoma except Hodgkin's lymphoma. His memorial plaque notes that he was also a reformer and a philanthropist.

Left: Plaque of William Marsden, 65 Lincoln Inn Fields, WC2A (Spudgun67). Right: Plaque of Sir Jonathan Hutchinson, 15 Cavendish Square, W1

William Marsden (1796–1867) studied surgery at St Bartholomew's Hospital. He is best known for having founded two hospitals. He was so disturbed by the difficulties of the poor to receive medical assistance that in 1828 he initiated a small dispensary from which what is known today as the Royal Free Hospital has developed. The unsatisfactory conditions of treatment of cancer patients prompted him in 1851 to establish another modest facility that is today the Royal Marsden Hospital.

Sir Jonathan Hutchinson (1828–1913) enrolled in St Bartholomew's Hospital to study medicine, and in 1850, he joined the Royal College of Surgeons. Surgery was his main calling but in addition he was much involved in scientific research and teaching. He practiced ophthalmology as well and was a skilled pathologist. He participated in official inquiries of vaccination in connection with such illnesses as smallpox and leprosy. He was a prolific author of research papers and a well-received lecturer. He was a great authority of syphilis and a number of other conditions that have been associated with his name ever since.

Plaques of Sir Edwin Saunders, 89 Wimbledon Parkside, SW19 (left), and Joseph Toynbee, 49 Wimbledon Parkside, SW19 (right, both, Spudgun67)

Sir Edwin Saunders (1814–1901) received his diploma from the Royal College of Surgeons in 1839. He then was employed by St Thomas' Hospital as a dental surgeon and lecturer in dental surgery. He researched and did surgery to repair cleft palate, a condition when the two plates of the skull that form the roof of the mouth are not completely joined. It was a great recognition of his professional excellence when in 1846 he was appointed to be Queen Victoria's personal dentist and dentist of the Royal Family.

Joseph Toynbee (1815–1866) completed his medical training at Hunterian Medical School and became an aural surgeon in St Mary's Hospital in Paddington. He researched the ear, gave and published lectures, and studied the anatomy of deafness. He and one of his sons, the social philosopher Arnold, share the blue plaque. Toynbee collected bone specimens from his numerous aural surgeries and he published a book about his studies. Upon his death, his collection was donated to the Hunterian Museum, but was destroyed in a bombing raid during WWII. Toynbee's premature death happened during his experimenting with a cure for tinnitus. It is a perception of noise or ringing in the ear, due to aging, an ear injury, or disorder in the circulatory system. Toynbee was testing whether inhaling the vapor of a combination of chloroform and hydrogen cyanide, HCN, might help when he was fatally poisoned by HCN. Toynbee involved himself in the local affairs in Wimbledon and the people there erected a memorial in his honor, a drinking fountain in the shape of a gothic tower, at the junction of Wimbledon Hill Road and High Street.

Plaque of Sir Frederick Treves, 6 Wimpole Street, W1, and his lithograph, 1884 (Wellcome Collection)

Earlier we mentioned Edward VII's appendectomy in 1902. It was the surgeon Sir Frederick Treves, eventually, Baronet Treves (1853–1923), who performed it. He studied medicine at London Hospital Medical College and qualified for membership in the Royal College of Surgeons in 1878. He became best known for performing appendectomies. He had an extraordinary appointment to Queen Victoria's Court. Once, Treves noticed a severely deformed man being exhibited by a showman in a place across the road from the London Hospital. He was James Merrick, known as the "Elephant Man." Treves freed the man from the showman and let him stay in the Hospital to the end of Merrick's life. Treves wrote a book about him and wrote other books, and not only on topics of medicine. He was a gifted and prolific author. His character also figured in fiction, Bernard Pomerance's play, and even an acclaimed film. A sad irony of fate was that the world's foremost expert of appendicitis died at the age of 70 from infection caused by a ruptured appendix.

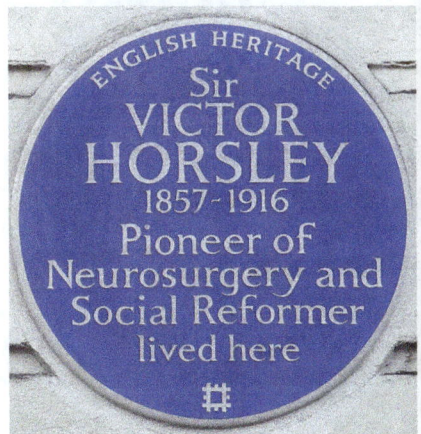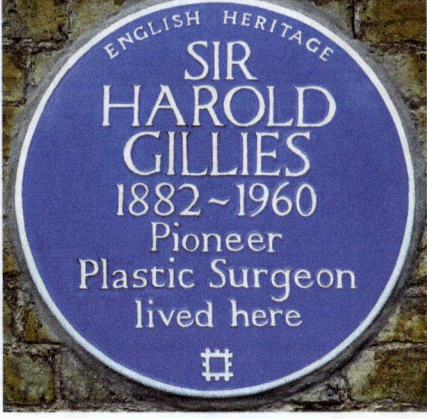

From left to right: Plaques of Sir Victor Horsley, 129 Gower Street, WC1; Sir Harold Gillies, 71 Frognal, NW3; and Sir Archibald McIndoe, 23–29 Draycott Avenue, SW3 (the right two, Spudgun67)

Sir Victor Horsley (1857–1916) studied medicine at UCL and in Berlin. He specialized in neurosurgery and was the first neurosurgeon appointed to what is today the National Hospital for Neurology and Neurosurgery. He was the first who removed a spinal tumor successfully. He researched the cerebral cortex in animal experiments and conducted other neurological investigations. He confirmed the efficacy of Louis Pasteur's anti-rabies vaccine. He was a recognized pathologist and founded the *Journal of Pathology*.

Sir Harold Gillies (1882–1960), often referred to as "the father of modern plastic surgery," was from New Zealand and studied at Cambridge University. His original specialization was in otolaryngology. During WWI and following some French examples, he initiated a facial injury ward at the Cambridge Military Hospital in Aldershof. Soon an entire new hospital was dedicated to facial repairs at Queen's Hospital in London. Over 5000 soldiers, many injured in the brutal trench warfare of WWI, were treated. In WWII, Gillies extended his services and he supervised and trained many plastic surgeons from the United Kingdom and the Commonwealth. In the years following the war, he was among the pioneers of sex reassignment surgery.

The plastic surgeon Sir Archibald McIndoe (1900–1960), a cousin of Gilles', grew up in New Zealand and studied medicine there. He learned pathological anatomy at the Mayo Clinic in the United States and moved to the United Kingdom in 1930. In London, he took up plastic surgery in a private practice, but then gained employment at the Hospital for Tropical Diseases and the London School of Hygiene & Tropical Medicine. In 1938, he became a consultant to the Royal Air Force (RAF), and this determined his service during World War II. He set up a unit for plastic surgery at the Queen Victoria Hospital in East Grinstead, Sussex. He turned out to be an extraordinarily successful reconstructive surgeon to the benefit of badly wounded and burned RAF pilots and other personnel. He introduced innovations such as reconstructing eyelids and fingers and was popular among his staff and patients. When he died, the much-decorated, but civilian McIndoe was accorded a resting place at the Central Church of the RAF of St Clement Danes, Strand, Temple, WC2R.

Plaques of Nicholas Culpeper, 92 Commercial Street, E1 (left), and Percy Lane Oliver, 5 Colyton Road, SE22 (right, both, Spudgun67)

Nicholas Culpeper (1616–1654) grew up fatherless, but thanks to his voracious reading in his grandfather's rich library, he developed a broad interest. He studied at Cambridge and became a botanist, physician, and astrologer. He questioned traditional medicine and determined to make medical care available to the masses. He operated a pharmacy and served as a field surgeon in military conflicts. His best-known activities were in "astrological botany" whereby he paired plants and diseases with consideration for planetary influence. He published his recommendations and they were widely read at the time.

Percy Lane Oliver (1878–1944) was not accepted for medical studies so he became a librarian, but his interest in and dedication to medicine never stopped. He was active in the local Red Cross movement. During WWI, he and his wife set up refugee hostels for which he received Royal recognition. In 1921, to respond to urgent need, they began organizing the first panels of voluntary blood donors. This eventually became what is today the National Blood Transfusion Service.

Wellcome Collection

From left to right: The Wellcome Collection at 183 Euston Road, NW1; Sir Henry Solomon Wellcome's portrait by Hugh Goldwin Riviere, 1906 (Wellcome Collection); and his plaque at 6 Gloucester Gate, NW1 (Spudgun67)

Two images from the Wellcome Collection. Left: Pharmacy in a painting, c. 1700, in France. Right: "The True Horrors of Cloning," a painting, 1998, by Heidi Kerrison

The Wellcome Collection, based on Sir Henry Solomon Wellcome's (1853–1936) art collection, opened in 2007. It is a combination of a museum and a library, aiming at presenting the interrelationship of health and humanity. It reflects Wellcome's dedication to human health and philanthropy. Silas M. Burroughs (1846–1895) and Wellcome founded the Burroughs Wellcome & Company in 1880, and it continues today in its successor institutions as well as in the Wellcome Trust, a large-scale medical charity. The Company in its time introduced innovations such as selling medicine in tablet form and used direct marketing to reach its customers. The Company took upon itself to engage in research in addition to production and trade, and established a number of research laboratories. In 1910, the American-born Wellcome became a British subject and over the years received a number of recognitions in his new homeland.

Innovators, Engineers, and Technologists

5

Iron sheet statue of Alan Turing at the southern end of St Mary's Terrace, W2

Allegorical representation of Engineering at the northwest corner of the Albert Memorial, Kensington Gardens, SW7

Examples of window keystone decorations on the back (Little George Street) of the former Middlesex Guildhall, now the Supreme Court building, Parliament Square, SW1. A man holding a pair of dividers and a cog (left) and a woman holding a sector gear and a pinion (right)

Allegorical engineer figures appear on the back of the guildhall building on Parliament Square, which is now the seat of the UK Supreme Court. The building was designed by James Glen S. Gibson and built in 1912–1913. We show two of the several similar window keystone decorations. They were created by Henry Charles Fehr, who was responsible for the many sculptures on the front façade of this building. The window decorations are of medieval style; the devices in the hands of the figures may be medieval or contemporary. Remarkably, both male and female engineers are depicted.

"Homage to Leonardo 'The Vitruvian Man'" by Enzo Plazzotta (completed posthumously by Mark Holloway, 1982), Belgrave Square, SW1

Plazzotta's sculpture is a reminder of Leonardo da Vinci's pen and ink drawing on paper, "The Vitruvian Man." Vitruvius was a Roman author, architect, and engineer in the first century BCE. His ideal was perfection of proportion.

Left: Leonardo da Vinci on the façade of the Royal Academy of Arts, Burlington Gardens, W1, from the court side, by Edward Bowring Stephens, 1874. Right: Leonardo da Vinci in the Frieze of the south side of the "Parnassus," Albert Memorial

The Renaissance Man from Italy, Leonardo da Vinci (1452–1519), usually, just Leonardo, was both a world-renowned artist and a brilliant inventor. He was also a mathematician, physicist, architect, anatomist, astronomer, geologist, and many more—a true polymath. His inventions as an engineer covered a similarly broad range, from flying machines, to solar power, even a double hull for building safer ships, just to mention a few. His statue stands at the Royal Academy of Arts, thereby referring to his creations in art, such as the *Mona Lisa* and *The Last Supper*. His statue is at the extreme left, followed by those of Flaxman, Raphael, Michelangelo, Titian, Reynolds, and Wren.

There is another Leonardo statue, one of the 169 full-size sculptures of the Frieze of Parnassus of the Albert Memorial (see Chap. 1). Leonardo is among the painters on the East side, standing next to two other giant representatives of the Italian Renaissance. Here he is represented for his innovations rather than for his paintings.

Electricity and Electronics

The subject matter of this section overlaps a great deal with physics. James Clerk Maxwell clearly manifests this overlap. He is mentioned among scientists (Chap. 3), but he was also the founder of electrical engineering. This section further demonstrates the symbiotic interconnection of electronics and telecommunications in that their progress acted as stimuli for each other.

Benjamin Franklin's plaques on the façade of 36 Craven Street, WC2N, and in the basement of the house (this image, courtesy of Matt Brown)

Benjamin Franklin (1706–1790) was born a British citizen in Boston and became an American statesman, one of the Founding Fathers of the United States. He was also a scientist and inventor. He showed the electrical nature of lightning in his famous experiment when in a storm he used a conducting kite to access current from a lightning bolt. The house at 36 Craven Street is the only former Franklin residence that still stands in London. It is now a museum, the Benjamin Franklin House. Franklin's inventions are many, chief among them a freestanding stove for home heating, bifocal eyeglasses, and the lightning rod. He has a number of memorials in New York and elsewhere.

Plaques of Colonel R. E. B. Crompton, 48 Kensington Court, W8 (left), and Alexander Muirhead, 20 Church Road, BR2 (both, Spudgun67)

Colonel R. E. B. Crompton (1845–1940) was an electrical engineer and both a scholar and entrepreneur. He received a classical education with extra mathematics, all the while tinkering with electrical devices. He served in the British Army, but in 1875 he decided on a commercial engineering career. He pioneered electric lighting and systems for supplying electricity. Later he was also active in the standardization of electrical systems and their extension internationally.

Alexander Muirhead (1848–1920) studied science at UCL and earned his doctorate in electricity at St Bartholomew's Hospital. This combination of technical skills and medicine served him well, as the first ever recording of a cardiogram is attributed to him. He designed a variety of precision instruments and introduced a number of innovations. He worked with the physicist Sir Oliver Lodge (1851–1940) on wireless telegraphy; the partners sold patents to Guglielmo Marconi.

Telecommunications

Tablet of Sir Francis Ronalds, 26 Upper Mall, W6 (left, Spudgun67), and his plaque, 1 Highbury Terrace, N5 (right, Matt Brown)

Sir Francis Ronalds (1788–1873) was an inventor and meteorologist. At the age of 28 he conceived a technique to transmit electrical signals over large distances. He conducted an experiment in his garden in which he sent electrical signals over 8 miles of insulated iron wire and thus anticipated the electrical age of mass communication. He offered his invention to the Admiralty, which showed no interest because it did not recognize its usefulness. Application of the electric telegraph was delayed, and it was only two decades later that it happened through the efforts of William Fothergill Cooke and Charles Wheatstone (see below). Ronalds was not discouraged by the lack of interest in his electric telegraphy, but went on inventing a host of other devices, including instruments for meteorology. Yet, he is best known as "the father of electric telegraphy."

Samuel F. B. Morse's plaque, 141 Cleveland Street, W1

Samuel F. B. Morse (1791–1872) was an American painter and inventor. When Joseph Henry developed an electromagnet, Morse conceived the idea of the electromagnetic telegraph. He designed a dot-and-dash code system, known as the Morse code for the new instrument. In his telegraph, he used a moving paper strip to record the messages received in his code. For quite some time, he continued his artistic activities and worked as an art professor at the University of the City of New York (as it was then) while moonlighting as an inventor. Eventually, he gave up his art and devoted himself fully to his technical activities. The blue plaque commemorates the period he lived in London where he studied painting and became a recognized portrait artist.

Plaque of Sir Charles Wheatstone, 19 Park Crescent, W1B (Spudgun67), and photograph of Sir William Fothergill Cooke (Wikipedia)

Sir Charles Wheatstone (1802–1875) was an inventor of broad scale in the fields of automatic transmission of electricity across distances and electrical measuring devices. He is best known for the Wheatstone bridge, an electrical circuit to measure electrical resistance. In his youth, he was interested in acoustics and the physical principles of musical instruments. This interest was the starting point of his leitmotiv of communicating over ever larger distances. His greatest achievements were in electricity and the electric telegraph in collaboration with Sir William Fothergill Cooke (1806–1879). Cooke studied medicine in Edinburgh, Paris, and Heidelberg and was a surgeon. But when in 1836 he saw a demonstration of an initial version of electric telegraphy, he gave up medicine and moved into engineering. Jointly, he and Wheatstone developed an electric telegraphic system for the railways that was instrumental for their modernization. As early as 1837 they filed for significant patents. Theirs was a fruitful joint venture. Although both received ample financial and societal recognition for their contributions, a petty priority dispute between them called for arbitration in which eminent engineer inventors represented each partner. Marc Isambard Brunel represented Wheatstone and John Frederic Daniell (of the Daniell cell, King's College) represented Cooke. The arbitrators concluded that the invention was a joint production.

Paul Julius Reuter's memorial by Michael Black, 1976, at the Royal Exchange, EC4

Paul Julius Reuter (1816–1899) was born in Kassel, Germany. From the start of his career, he was interested in communications and how to make them faster. He founded a pigeon post service between the northwestern German town Aachen and Brussels. His winged couriers proved faster than the trains used at the time for carrying dispatches. When he learned about the possibilities of telegraph services, in 1851 he moved to London and opened an office of telegraphic communications at the Royal Exchange to transmit stock prices. Reuter's bust was unveiled on the 125th anniversary of the opening of this office. Eventually, the Reuter News Agency formed, which is today part of the Thompson Reuter Corporation.

Lord Kelvin, left: Bust by Archibald MacFarlane Shannan at the Science Museum. Right: Photograph by Harry Herman Salomon (Wellcome Collection)

Lord Kelvin (William Thomson, 1824–1907) has already figured among the scientists (Chap. 3). Here we remember him as an engineer and inventor, especially for his contribution to the electric telegraph. He was a major participant in establishing the underwater cable across the Atlantic Ocean, which brought him fame and wealth. He then took part in endeavors of laying cables elsewhere. He was one of the four inaugural scientists appointed on June 26, 1902, for membership in the newly established Order of Merit (OM), which is limited to 24 and is personally selected by the British monarch. It may be the most prestigious honor on Earth.[1]

---

[1] The other three inaugural members of the OM were the physicist Lord Rayleigh (Chap. 3), the astronomer Sir William Huggins, and the surgeon Lord Lister (Chap. 4). Sir William Huggins (1824–1910) and his wife Margaret Lindsay Huggins (1848–1915) pioneered the use of spectroscopy in astronomy. The two worked together for 35 years as equal partners. He received many distinctions and awards. It was seen as a major achievement when her name was mentioned as a co-author of a joint paper.

Plaques of David Edward Hughes, 94 Great Portland Street, W1W (left), and Sir Ambrose Fleming, 9 Clifton Gardens, W9 (right)

David Edward Hughes (1831–1900) was born in Britain and moved with his family to America. He excelled in music and had a professorship in Kentucky. By the time he was 25 years old, he had made significant inventions and moved back to Britain to utilize them. His printing telegraph was a huge success and facilitated the spread of telegraphing. The Western Union Telegraph Company thrived on using it and so did international companies. He then prototyped the microphone, followed by other inventions. We have no information about his educational background, but he was a true genius. Also, he was modest, published little about his inventions, and readily recognized others, such as Heinrich Hertz and Guglielmo Marconi, for inventions similar to his. The Royal Society recognized his merits and founded the Hughes Medal, for which J. J. Thompson was the first recipient, followed by many big names in physics and invention.

Sir Ambrose Fleming (1849–1945) wanted to become an engineer from early childhood, but could reach his goal only by alternating between work and learning. He studied at UCL, Cambridge University, and what is today Imperial College. James Clerk Maxwell was one of his professors. Fleming taught at various schools, including UCL, and worked for various companies, among them the Edison Electrical Light Company and Guglielmo Marconi's company. Fleming remained bitter about his employment by Marconi, for he received neither fair payment for his work nor the promised acknowledgment for his intellectual contribution to Marconi's major achievement of the time: the first radio transmission across the Atlantic Ocean. Many consider Fleming's invention of the vacuum tube the dawn of the era of electronics. He had a number of other inventions in photometry, radio technique, and elsewhere in electronics.

Plaque of Hertha Ayrton, 41 Norfolk Square, W2

Hertha Ayrton (1854–1923) was a physicist, engineer, and inventor. She was born as Phoebe Sarah Marks. Her father was a Polish Jewish immigrant who died early. Poverty made her childhood difficult, but her talent for science and mathematics showed early, so she received support from family and friends and attended Girton College in Cambridge. While yet a student, she constructed an apparatus for measuring blood pressure. In 1880, she passed the necessary exams to study for a degree at Cambridge, which would not award full degrees to women at the time, so she moved to London. She received her Bachelor of Science degree in 1881 from the University of London. In 1884, she filed for her first patent, a line divider, which was a mechanical instrument capable of dividing a line into any number of equal parts and enlarging and reducing figures. She successfully registered 25 more patents during her life. Most of her inventions and patents concerned arc lamps and electrodes. This was after she had extended her studies to electrical engineering. In 1885, she married her former UCL professor of electricity, William Edward Ayrton. In 1900, she was nominated to be elected Fellow in the Royal Society. Until then there had not been any female member, and she was not to be either, a denial based on curious reasoning. The committee observed that she was married, and her husband was already a Fellow. Because under British law husband and wife were considered to be one person, she could not be a member because she did not represent a separate person.[2] Though shut out of the Royal Society, she became the first woman awarded a prize by the Royal Society—she received the Hughes Medal in 1906 for her achievements in studying the electric arc and the motion of ripples in sand and water. Understandably, she was active in politics and fought for strengthening women's positions in scientific societies and elsewhere. In 2010, a panel of experts of the Royal Society named her as one of the ten most influential women scientists in British history.[3]

Plaques of David Gestetner, 124 Highbury New Park, N5, and Frederick George Creed, 20 Outram Road, CR0 (both, Spudgun67)

David Gestetner (1854–1939) was Jewish-Hungarian-born engineer who left Hungary first for Vienna, then America, then eventually moved to London. By then, he had conceived the idea of a duplicating method, which originated from an office accident. He spilled ink on a pile of papers and observed that the same pattern repeated throughout the entire pile. From this observation, he developed his stencil duplicator, which quickly became an international success. This was well before photocopying and the possibility of mass producing copies in an inexpensive way transformed modern life.

Frederick George Creed (1871–1957) was an electrical engineer who invented the teleprinter. He was born in Canada and operated first in Glasgow, then moved operations to London. The newspaper industry first recognized the utility of the device.

---

[2] The crystallographer Kathleen Lonsdale and biochemist Marjory Stephenson became the first female Fellows in 1945.

[3] The full list was Caroline Herschel (astronomer), Mary Somerville (physicist), Mary Anning (paleontologist), Elizabeth Garrett Anderson (physician), Hertha Ayrton (physicist), Kathleen Lonsdale (crystallographer), Elsie Widdowson (nutritionist), Dorothy Hodgkin (crystallographer), Rosalind Franklin (biophysicist), and Anne McLaren (geneticist).

  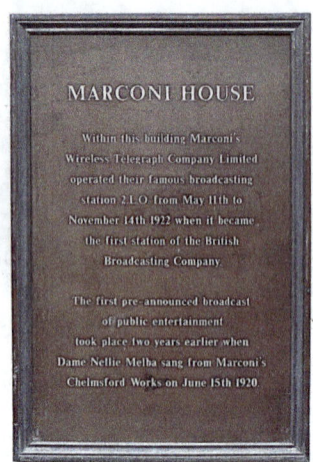

From left to right: Plaques of Guglielmo Marconi, 71 Hereford Road, W2 (Spudgun67), and on the façade of British Telecom Centre, 81 Newgate Street, EC1A, and of Marconi House on its façade, 335 Strand Underpass, WC2R

From left to right: Bush House (now part of the Strand Campus of King's College), 30 Aldwych, WC2; BBC memorial plaque on its façade; and plaque of BBC School Radio, 1 Portland Place, W1

A few tidbits referring to telecommunications and the contributions of Guglielmo Marconi (1874–1937) are presented here. Marconi was an Italian electrical engineer and world-famous inventor. He is best known for his discoveries that were fundamental to make long-distance radio transmission possible. He and Karl Ferdinand Braun (1850–1918) jointly received the 1909 Nobel Prize in Physics "in recognition of their contributions to the development of wireless telegraphy."

The current headquarters of British Telecom (BT) was the venue from where Marconi "made the first public transmission of wireless signals on 27 July 1896." The posters in its lobby (as of spring 2019) narrate some focal points in British telecommunications. One of them commemorates the birth of British TV in 1927. It mentions John Logie Baird (see below) with emphasis of his contribution to the development of television. Baird carried out his early broadcast from London to Glasgow over telephone lines.

According to the BBC memorial plaque on the façade of Bush House "International radio, television and online content [was] made here 1941–2012." Now this building is part of the Strand Campus of King's College. British broadcasting specifically for school children commenced in 1924. The BBC School Radio operated at 1 Portland Place, W1, between 1952 and 1993.

From left to right: Plaques at John Logie Baird's former residence, 3 Crescent Wood Road, SE26 (Spudgun67); for his first demonstration of television, 22 Frith Street, W1; and of the world's first high-definition television service at Alexander Palace ("Ally Pally"), Wood Green, N22 (Matt Brown)

John Logie Baird (1888–1946), a pioneer of television, studied at the Glasgow and West of Scotland Technical College (as it then was) and the University of Glasgow. He was among the most prominent contributors to the invention of television.[4] He has blue plaques in London. One of them commemorates Baird's first demonstration of his television before some members of the Royal Institution at his then home, in Frith Street. There is another one at another of his domiciles, Crescent Wood Road. Yet another one (not shown here) was erected at the site where Baird broadcast the first television program in Britain on September 30, 1929, in Long Acre. The next advance was the commencement of the world's first regular high-definition television service in 1936, which is commemorated by yet another plaque.

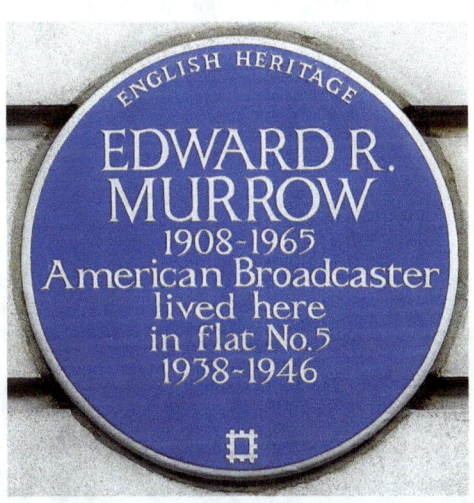

The legendary broadcaster Edward Murrow's plaque, Weymouth House, 84–94 Hallam Street, W1W

Edward Roscoe Murrow (1908–1965) was an American broadcaster and war correspondent for the CBS network. He lived in London 1938–1946. For more than 2 years Britain was already engaged in WWII while the United States was still neutral. It was of utmost importance for Britain to keep the American people informed about the ordeal of the British people, especially during the *Blitz*. Murrow's daily program, beginning with the catch phrase "*This* is London" and ending with "Good Night, and Good Luck," fulfilled this need in an exemplary way. His reporting made broadcast history.

---

[4] See also the memorials of Vladimir K. Zworykin in Moscow and Philo T. Farnsworth in Washington, DC; I. Hargittai and M. Hargittai, *Science in Moscow: Memorials of a Research Empire* (Singapore: World Scientific, 2019).

Computing

From left to right: Charles Babbage's plaque at 1A Dorset Street, W1; his wood engraving after T. D. Scott, 1871 (Wellcome Collection); and another plaque at the corner of Walworth Road and Larcom Street, SE17 (Matt Brown)

Model of Babbage's difference engine No. 2 at the Science Museum. He designed the original in 1847–1849, and the Museum had the model built in 1985–2002

Charles Babbage (1791–1871) was an English mathematician and inventor. He studied at Trinity College, then at Peterhouse, of Cambridge University. To augment the curriculum, he and his friends, John Herschel and George Peacock, also future renowned scientists, formed a private circle, the Analytical Society, to advance their studies. After leaving university, Babbage lectured at the Royal Institution and was elected Fellow of the Royal Society. Between 1828 and 1839, he occupied the prestigious Lucasian Professorship of Mathematics in Cambridge. His career was unusual; he tried his talent at a variety of applications and was embroiled in controversies. He did pioneering work in metrology and involved himself in innovative engineering projects, in part, through his interaction with Isambard Kingdom Brunel. What happened to his ophthalmoscope may have been a typical fate of his initiatives: the experts ignored it and it was later independently invented by Hermann von Helmholtz. Babbage made significant inroads into cryptology.

Of Babbage's many contributions, his pioneering works in computing stand out. He built some of the first mechanical computers. His difference engine computed values of polynomial functions. The name of the machine indicated that its operations were based on finite differences; in other words, they were limited to addition and subtraction. Babbage struggled with finding adequate funding for his projects, so important designs remained unfinished. His analytical engine was a more sophisticated computer representing a transition from mechanized arithmetic to a fuller-scale computation. It was to be operated with punch cards. Its ever-improving designs were gradually getting closer to the ideas of the modern computer.

Ada Lovelace's plaque, 12 St James's Square, SW1, and her portrait about 1840 by Alfred Edward Chalon (Wikimedia)

Ada Lovelace (née Byron, 1815–1852) was an English mathematician, the daughter of the poet Lord Byron. He died when she was 8 years old. Her mother, Anne Isabella Milbanke, encouraged her to study mathematics and develop interests far from Byron's. Ada became Countess Lovelace on account of her mother's second husband. As a teenager, Ada met Charles Babbage through her mentor, Mary Somerville. Ada devoted herself to working on the analytical engine. Her notes are considered to be the first attempt to create a computer program. In this, her interest superseded Babbage's. Furthermore, her notes revealed her anticipation of the importance of the computer in the lives of individuals and society alike. Her early death has been attributed to cancer and to medical malpractice. The significance of her contribution to the development of the computer has been debated, but there is consensus that it had been underestimated. *The New York Times* published her obituary 166 years after her death.[5]

---

[5] Clare Cain Miller, "Ada Lovelace: A Mathematician Who Wrote the First Computer Program." *The New York Times*, March 8, 2018.

From left to right: Alan Turing's portrait (US NSA); his plaque, 2 Warrington Crescent, W9, London; and his statue by Glyn Hughes, 2001, in Sackville Park, Manchester

Alan M. Turing (1912–1954) was an English mathematician and computer pioneer. He is best known as a code breaker in World War II and as one of the fathers of the modern computer. From early age he displayed exceptional talent and fierce interest in science and mathematics. He started his undergraduate studies at King's College of the University of Cambridge in 1931 and he became a Fellow in 1935. In 1936, he published a seminal paper whose essence was that a machine could be built to compute anything a human could. This was the idea of software, and in way of tribute, such machines have been called Turing machines. Afterward, Turing spent 2 years at the Institute for Advanced Study (IAS) in Princeton and earned his PhD degree there. The Princeton scientist John von Neumann, another pioneer of the modern computer, wanted Turing to stay at Princeton, but Turing returned home. To what extent their discussions advanced their thinking about programmable computers will never be known. We suggest a possible formulation for their respective contributions in that Turing originated the idea of the programmable computer and von Neumann built the first one with programs stored in its memory.

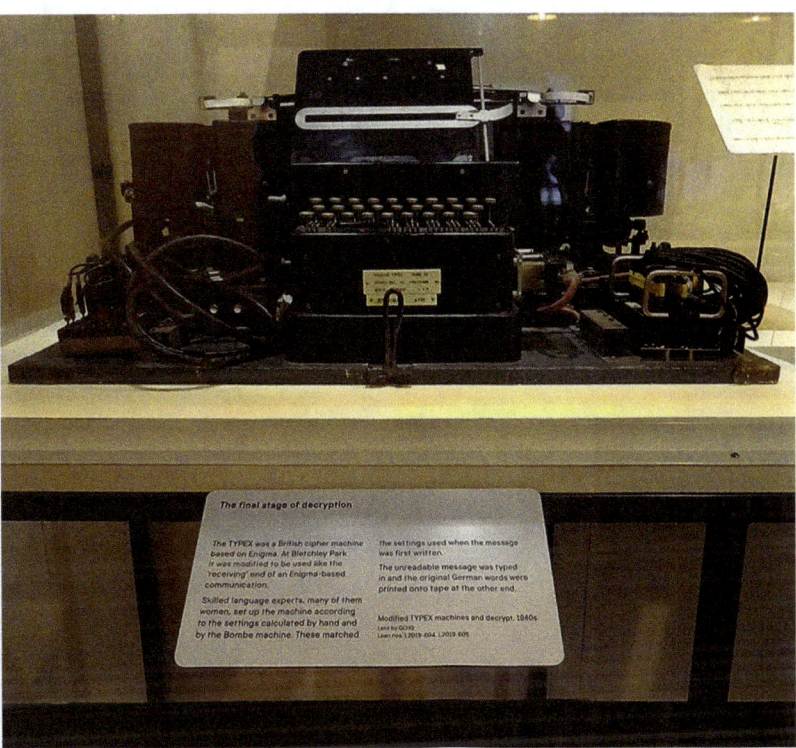

Left: Enigma machine exhibited at the British Library in front of the entrance to its Turing Institute. The Enigma was manufactured in Germany in 1944 for the encryption of communications between the U-boat division of the German Navy and its naval bases. Right: The Typex British cipher machine based on the German Enigma is at the Science Museum

Upon his return from America to England, Turing attended Ludwig Wittgenstein's lectures in Cambridge and engaged in discussions with the professor. From 1938, first as a part-time associate, soon full-time, Turing became member of the Government Code and Cypher School (GC/CS) at Bletchley Park. He wrote up his mathematical advances in cryptography in two papers; they were declassified only 70 years later, in 2012. He designed an electromechanical machine, the "bombe," to break the German Enigma code. The benefits in saving lives and materiel due to Turing and his colleagues' achievements at Bletchley Park cannot be overestimated.

After the war, Turing's activities focused on the stored-program computer. He lived and worked in London between 1945 and 1947. He then returned to Cambridge, followed by his appointment to the Mathematics Department of Victoria University of Manchester. In 1950, he published another seminal paper, "Computing Machines and Intelligence." In it, he focused on the question whether a machine can think. He proposed a test to determine just that and called it "The Imitation Game." This was to become the title of the film about his life in 2014.

Turing had a broad interest in applying his talent in mathematics, mathematical logic, and computation. In 1951, he became interested in the origin of biological form. In 1952, he published another important study, "The Chemical Basis of Morphogenesis." He pointed to the importance of chemical reactions and the diffusion of the reaction products in the formation of characteristic patterns and shapes of living organisms.

Turing's exceptionally creative activities were destroyed when his homosexuality—a criminal offense at the time—came to light and he was convicted. He was given a choice between imprisonment and hormonal treatment. He chose the latter. His conviction had additional consequences. His security clearance was revoked in Britain, and he was banned from the United States where he had established fruitful interactions. Turing committed suicide by swallowing cyanide in 1954 although this has been disputed and there have been suggestions that what happened was accidental poisoning.

For decades, Turing remained under the cloud of disgrace. In 2009, Prime Minister Gordon Brown apologized for Turing's "appalling" treatment. However, an official pardon was not forthcoming until Elizabeth II exercised the "royal prerogative of mercy." In 2014, Turing was inducted into the National Security Agency/Central Security Service (NSA/CSS) Cryptology Hall of Honor at the National Cryptology Museum (NCM) in Annapolis Junction, MD, USA. He was cited as follows: **"Dr. Alan Turing:** A brilliant theoretician whose concepts underpin 70 years of computing, enabling processing of very high-grade enciphered communications and led to development of the modern computer, and whose work turned sophisticated encrypted messages into actionable intelligence."[6] Only in 2017 was his amnesty recognized by law, a law known popularly as the "Alan Turing law," retroactively pardoning those who had been victimized for homosexual acts. The tablet at the Turing statue in Manchester stated as early as 2001: "Father of Computer Science/Mathematician, Logician/Wartime Codebreaker/Victim of Prejudice."

Turing memorial adjacent to the canal entrance to Paddington Underground beneath the Bishop Road Bridge

In 2012, an unusual memorial of Turing's legacy was unveiled beneath the Bishop Road Bridge flyover adjacent to Paddington Station. Arrays of light-emitting diodes (LED) flash up displaying words from a poem by the poet Nick Drake, referring to Turing. The installation resembles the displays of railway stations communicating the ever-changing information about the arrivals and departures of trains. The words of the poem in this dynamic format keep changing although few of the passing-by crowd seem to notice or read the tablet of description of the memorial.

---

[6] https://www.nsa.gov/about/cryptologic-heritage/historical-figures-publications/hall-of-honor/2014/aturing.shtml

Steel statues of three local heroes at the southern end of St Mary's Terrace, W2: from left to right, Mary Seacole, Alan Turing, and Michael Bond

The steel statue of Turing, used as the frontispiece of this chapter, is one of three statues honoring local heroes. The memorial was installed in the framework of Sustrans, a sustainable transport charity, aiming at creating a nationwide network of walking and cycling routes. Beside Turing, the other two are Nurse Mary Seacole (see more about her in Chap. 4) and Michael Bond, the author and creator of the popular children's character Paddington Bear.

Royal School of Mines (Imperial College)

Royal School of Mines, Imperial College, Prince Consort Road, SW7, corner of Exhibition Road and Prince Consort Road (left) and its entrance on Prince Consort Road (right)

The Royal School of Mines—today, the Department of Earth Sciences and Engineering, comprising Geology, Geophysics, Mining and Petroleum Engineering—is part of Imperial College. Its magnificent building has several memorials.

Busts of Alfred Beit (left) and Julius Wernher (right) by Paul Montford flank the main entrance to the Royal School of Mines

Alfred Beit (1853–1906) and Julius Wernher (1850–1912) were German-born British diamond and gold entrepreneurs who amassed great wealth in South Africa. They and the trusts they established donated significant amounts of money to support institutions of culture and higher education, among them the Royal School of Mines.

Bust of William Smith at the Royal School of Mines and his plaque, 15 Buckingham Street, WC2

William Smith (1769–1839), geologist, was self-educated and started his career as a surveyor. His initial appointment was by John Rennie. When Smith inspected coal mines, he noted the variety of layers of rock and coal. Eventually, he developed his theories of stratigraphy. He created and published the first geological map of Great Britain and earned the popular title, "Father of English Geology." An identical bust with the one at the Royal School of Mines is at the Oxford University Museum of Natural History.

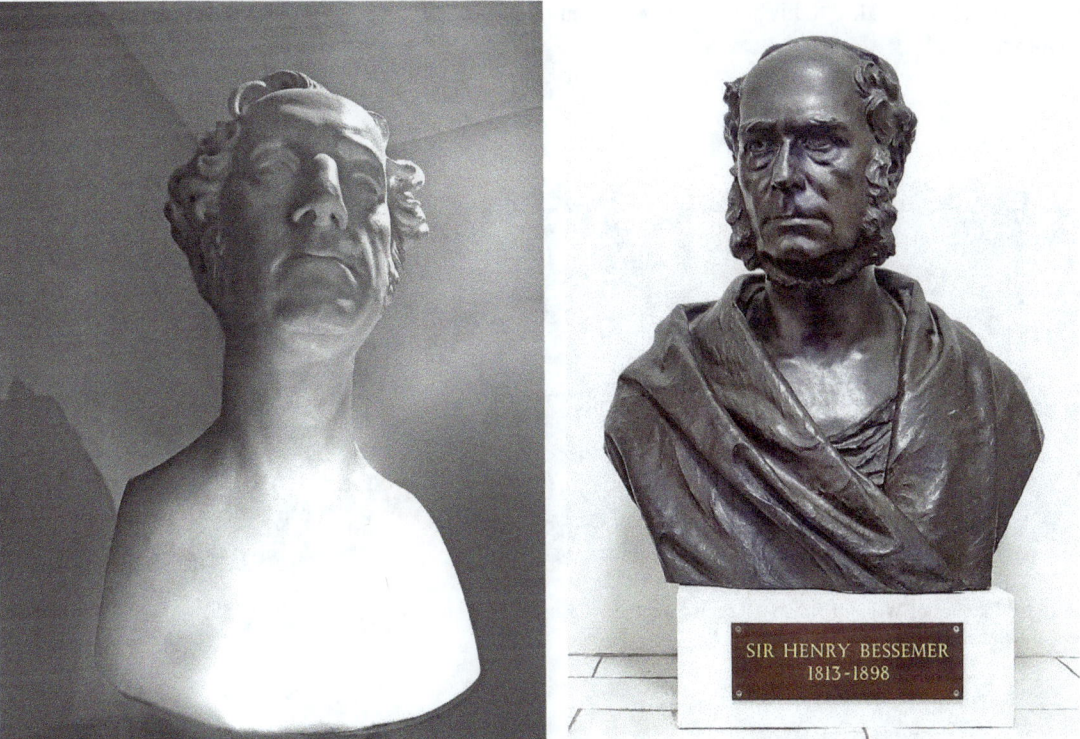

Busts of Henry de la Beche (left) and Henry Bessemer (right) at the Royal School of Mines

Geology, Geodesy, Geography

Henry de la Beche (1796–1855) was a geologist and paleontologist. He was interested in geology from early childhood and was a friend of the paleontologist Mary Anning. He joined the Geological Society of London in 1815, founded the Geological Survey of Great Britain in 1835, and served as its first director. In 1841, he established the Museum of Economic Geology, which is now the Earth Galleries of the Natural History Museum.

Henry Bessemer (1813–1898) pioneered the production of what has become known as Bessemer steel. The problem with steel had been the uneven carbon content of iron, making it brittle. His revolutionary procedure blew air through the molten iron: the hot air oxidized carbon along with other impurities, and the volatile products left the molten iron. This was not Bessemer's first invention, nor was it his last. He was prolific and owned well over 100 patents. In the history of inventions, patent infringements and other disputes and trials were rather common. It is refreshing then that Bessemer's career offers a different example. A Scottish engineer, James Nasmyth (1808–1890), had also been working on a procedure for improving steel production and he had achieved promising results. However, when he heard about Bessemer's account of the successful new technology, he abandoned his own project. When Bessemer learned about this, he offered Nasmyth one-third share of his own patent as a recognition of the other inventor's efforts. Nasmyth declined the generous offer. Nasmyth was about to retire although he was not yet even 50 years old, but he could afford devoting the rest of his life to his hobbies, viz., astronomy and photography.

Allegorical representation of Geology at the northwest corner of the Albert Memorial

  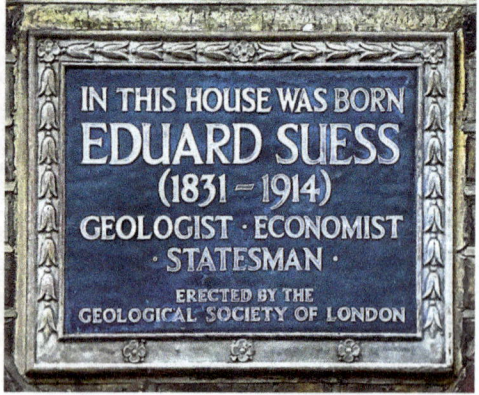

From left to right: Plaques of William Roy, 10 Argyll Street; Sir Charles Lyell and the statesman W. E. Gladstone, 73 Harley Street, W1; and Eduard Suess, 4 Duncan Terrace, N1

William Roy (1726–1790) was an engineer and military officer who used mathematical methods and modern technology for geodesic surveying. As a result, alas, posthumously, *The Ordnance Survey of Great Britain* was created in 1791. The memorial plaque is on the façade of the house where he lived.

Sir Charles Lyell (1797–1875) started in the legal profession but switched to geology, thus influencing the paradigm shifts of modern biology. His areas of inquiry included climate change, earthquakes, and various periods of the history of Earth. He was a friend of Charles Darwin and assisted the publication of Darwin's and Alfred Russel Wallace's papers in 1858. He did so despite being religious and having reservations about the theory of natural selection advanced in these treatises. Lyell authored the monograph *Principle of Geology* and served as President of the Geological Society. He is buried in the north aisle of the nave, Westminster Abbey. There is a marble bust by William Theed, Jr., on the window ledge and a gravestone on the floor. It reads, among others, "Throughout a long and laborious life he sought the means of deciphering the fragmentary records of the Earth's history in the patient investigation of the present order of Nature enlarging the boundaries of knowledge and leaving on scientific thought an enduring influence."

Eduard Suess (1831–1914) was an Austrian geologist and economist. He was born in London and his birthplace is commemorated with a plaque by the Geological Society of London. The family moved to Prague, then, to Vienna and he completed his schooling there. He was appointed Professor of Paleontology at the University of Vienna in 1856 and Professor of Geology in 1861. He had interesting and comprehensive ideas about the development of the continents of Europe and Africa and the origin of the Alps and the Mediterranean Sea. He had international fame: his busts stand in Vienna on Schwarzenberg Square, erected in 1928, and in Moscow in the Earth Science Museum of Lomonosov University, erected around 1950.

Left and middle: Harry Beck's plaques at his birthplace, 14 Wesley Road, E10 (left, Spudgun67), and where he used to board the tube, the southbound platform of Finchley Central Station, Northern Line, N2 (Matt Brown). Right: Phyllis Pearsall's steel sheet statue, Brunswick Quay at Greenland Dock, Rotherhithe, SE16

Harry Beck (1902–1974) was a technical draughtsman who created the original version of what is now the world-famous London underground map. The scheme has become a design classic, but initially the authorities declined its acceptance as they found it too radical. However, when copies of a trial print-run had become available, the public became enthusiastic about it.

Phyllis Pearsall (1906–1996) was a painter and writer. She produced the useful A–Z street map of London and founded the Geographers' A–Z Map Company. Her steel sheet statue is one in a group of three: the other two depicting the actor Sir Michael Caine and the cyclist Barry Mason.

Chemical Engineering (ICI—Imperial Chemical Industries)

Stone portraits on the façade of the former ICI building (see, p. 102). From left to right: Harry McGowan, Alfred Nobel, Ludwig Mond, and Alfred Mond

Alfred Nobel (1833–1896) was a Swedish chemical engineer, manufacturer, and inventor. He lived in Sweden, Russia, France, and, again, in Sweden. His most famous invention was the explosive dynamite—a stabilized form of nitroglycerin. Later, he invented smokeless gunpowder and then gelignite, another explosive, known as blasting gelation. His inventions were most lucrative and he built up an international empire of business and industry. Drawing upon his enormous wealth, his Will established an award, which became the Nobel Prize, in five categories, viz., physics, chemistry, physiology or medicine, literature, and peace. The Nobel Prize has become the most prestigious and famous international recognition.

Ludwig Mond (1839–1909) was a German-born British chemical engineer, a disciple of the renowned German chemist Robert W. Bunsen. The elder Mond was an innovator and discoverer in chemistry. He developed Ernest Solvay's process of producing soda into a large-scale industrial process. He discovered the compound nickel carbonyl, which led him to work out nickel purification—the Mond process of great industrial significance. He built up a successful business in Great Britain. He sponsored the founding of the Davy-Faraday Laboratory at the Royal Institution where a medal with his portrait commemorates him.

Alfred Mond (1868–1930) was a British industrialist, Ludwig Mond's son. Alfred helped form ICI and served as its first director, but he was more a politician than a manufacturer.

Harry (Henry) McGowan (1874–1961) was a chemical manufacturer. He started his career in Alfred Nobel's employment and became an advocate of joining major chemical companies into one big entity. He was instrumental in forming Imperial Chemical Industries, which he directed between 1930 and 1950.

Biotechnology

Chaim Weizmann's memorial plaque, 67 Addison Road, W14, celebrates the scientist and Israel's first president

Chaim Weizmann (1874–1952) was a biochemist and a Jewish statesman. Born in Russia, he studied in Switzerland and Germany, then became a Zionist leader in England. His achievements in chemistry helped the British war efforts in WWI. This made it possible for him to have a voice in the negotiations leading to the Balfour Declaration in 1917 committing Britain to the establishment of a Jewish state in Palestine. Weizmann helped found the Hebrew University of

Jerusalem, the Daniel Sieff Research Institute (today, the Weizmann Institute), the Jewish Brigade in World War II, and the recognition of the new Jewish state—Israel. He was Israel's first president.

Weizmann's chemistry helped to create a synthetic alternative to rubber. He used bacteria in a fermentation process on an industrial scale to generate acetone and butyl alcohol. Acetone was necessary to produce various explosives, and butyl alcohol provided the starting materials for the production of many organic substances of industrial importance. According to many, Chaim Weizmann was the father of biotechnology.

Civil Engineering

Much of what civil engineering does is directly beneficial to human life—lifting the quality of life—in particular in big cities. Some of it is in tight coordination with public health. We have seen in Chap. 4 that John Snow uncovered issues in sanitation and removed the handle of a pump, thereby preventing people in the neighborhood from consuming contaminated water. However, it was only the diagnosis of the problem. The cure had to come from civil engineering. Joseph Bazalgette provided it when he designed an improved sewage system and drainage (see below).

The Elder John Rennie's bust by John Ravera, 1992, at Spirit Quay, off Vaughan Way, overlooking the London Docks in Wapping, E1 (left), and the bust of James Walker by Michael Rizzello, 1990, Brunswick Quay at Greenland Dock, Rotherhithe, SE16 (right)

John Rennie (the Elder, 1761–1821) was a civil engineer and technical innovator, who designed docks, canals, and bridges, including the Southwark, Waterloo, and London Bridges, and improved drainage. He is buried in St Paul's Cathedral.

James Walker (1781–1862) was a civil engineer who designed docks, bridges, lighthouses, railways, and other structures. He studied at Glasgow University and apprenticed in London. Eventually he and his partners had their own engineering firm. He was a major figure in the British engineering community and the President of the Institution of Civil Engineers, 1834–1845. This organization commissioned his bust shown above.

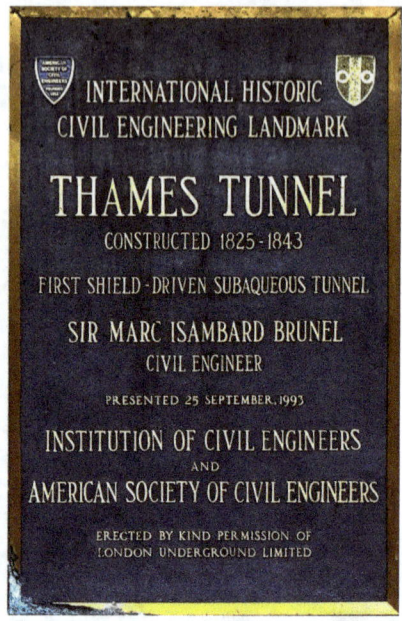

The Rotherhithe Overground Station and a memorial plaque at the Station

The Brunel Museum and a memorial plaque on its façade, Railway Avenue, Rotherhithe, SE16

In the Rotherhithe Station of the London Overground is a memorial plaque (1993) designating the Thames Tunnel, constructed and built between 1825 and 1843, as an International Historic Civil Engineering Landmark. It was the "first shield-driven subaqueous tunnel." What is presently the Brunel Museum used to be the Engine House of the Thames Tunnel, designed by Sir Marc Isambard Brunel (1769–1849), just behind the Rotherhithe Station. It has been a museum since 1961 and was given the name Brunel Museum in 2006.

   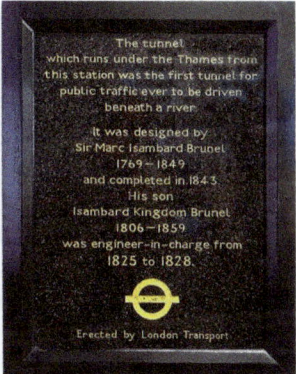

From left to right: Plaque of the two Brunels, 98 Cheyne Walk, SW10 (Spudgun67); busts of Sir Marc Isambard (left) and Isambard Kingdom (right) in the Brunel Museum; and tablet at Wapping station

Sir Marc Isambard Brunel was born in France but fled to the United States during the French Revolution. There, he held the position of Chief Engineer of New York City. In 1799 he moved to England to work on a Royal Navy project. Later, he was engaged in other government constructions and his most famous design was the tunnel under the Thames. He patented a tunneling shield whose technology became much utilized afterward in the tunnels of the London Underground and elsewhere. The Thames Tunnel between Rotherhithe and Wapping was opened in 1843, and on July 26, Queen Victoria and Prince Albert visited it. Prince Albert had been a keen supporter of the project. During the first decades the Tunnel was used for pedestrian crossing; the first train passed through it in 1869. Today it is used by the East London Line of the London Overground.

Statues of Isambard Kingdom Brunel, from left to right: by Carlo (Charles) Marochetti, 1877, 12 Temple Place, WC2R, at the Brunel Museum; and by John Doubleday, 1982, at Euston Station, WC1

There are at least four statues of Isambard Kingdom Brunel (1806–1859) in London, a degree of recognition unusual for a scientist or engineer. In addition to the three shown here, there is one by Anthony Stones (2006) on the campus of Brunel University, named after him, in Uxbridge, UB8. His unusual name requires comment. His father, Marc, was French, his mother English. His given name, Isambard, is Norman-German, meaning "iron-bright." His middle name was his mother's maiden name. Initially, he was assisting his father, who was the chief engineer of the construction of a tunnel under the River Thames in London. Isambard Kingdom Brunel was eventually the engineer directing the building of a number of important bridges in London and elsewhere. He was also involved in the construction of railway lines at the time of booming expansion of the British railway system. The construction of new railway lines often included the building of new bridges along the route, of which he was also the undisputed master. He was the chief engineer of the Great Western Railway from London to Bristol (later to Exeter). Paddington Station, its London terminus, was designed by him. His solutions were often unusual, sometimes quite daring. He had great imagination and envisioned, for example, the London-to-New York connection as one continuous trip. Always ready to come up with solutions to current problems where engineering could be of assistance, he became involved with developing and modernizing transatlantic shipping. Another case in point was his design of an army hospital during the Crimean War. He managed this task splendidly although he had not had any experience with medical institutions in his previous work. Brunel's fame has remained legendary, and numerous memorials pay tribute to him far beyond London.

Left: Memorial of Sir Joseph W. Bazalgette by George Blackall Simonds (1899) on Victoria Embankment, WC2. The inscription reads: "Sir Joseph Bazalgette CB, Engineer of the London Main Drainage System and of This Embankment." Right: Memorial at the Chelsea Embankment (1874), designed by Sir Joseph Bazalgette

Left: Cartoon of Sir Joseph William Bazalgette with his design of the modern sewerage system and embankments in London (Wellcome Collection). His plaque, 17 Hamilton Terrace, NW8 (Spudgun67)

Sir Joseph W. Bazalgette (1819–1891) was a civil engineer who gained experience in land drainage, land reclamation, and the expansion of the railway network. His most important commission was his appointment to be the Chief Engineer of the Metropolitan Board of Works of London. Subsequent cholera epidemics, among them the outbreak of 1853–1854, killed thousands of Londoners. Then, in July–August, the weather brought hot air to London and the smell became unbearable—a thermal inversion known as the Great Stink. The so-called miasma, the foul air from human waste and industrial effluent, was considered to be the culprit, the cause of both the epidemics and the stink. In contrast, and in concert with John Snow's solid findings (Chap. 4), Bazalgette came to the conclusion that the contaminated water was the cause of the epidemics and the stink. The work on the new sewage system started in 1859, and by the mid-1866 most of London (though not yet East London) enjoyed better health brought by the modern system. When the old problems re-appeared in East London, even the most stubborn advocates of the miasma concept became convinced that Snow and Bazalgette were right. The system was completed by 1875.

From left to right: Engine house at the Crossness Pumping Station, wood engraving by John Mayhew Williams, 1865 (Wellcome Collection); detail of Bazalgette's bust at the Crossness Pumping Station; and an image from the interior of the Station (the two latter, Matt Brown)

The development of sewage treatment facilities continued after Bazalgette's passing, but many think he deserves greater recognition. His memorial carries the Latin motto, FLUMINI VINCULA POSUIT, "He placed the river in chains (or "He placed chains on the river"). It is an impressive memorial, but considering the value of his deeds and the number of lives he must have saved, the frequent lamentation that he would deserve a larger memorial is understandable. His foresight can be seen in choosing the diameters of the sewers; he clearly had a vision of the future. As a consequence of his activities, the cholera epidemics as well as the typhoid epidemics have been eliminated. For proper drainage, pumping was necessary. The pumping stations he designed were not only efficient but also beautiful due to their exquisite decoration. They have become items of cultural heritage, especially Abbey Mills in Stratford and Crossness on the Erith Marshes. The sewers ran in new embankments, the Victoria, Chelsea, and Albert Embankments, that created space for plenty of parks—they, too, are a manifestation of usefulness and beauty.

The Crossness Pumping Station is especially rich in decoration and has been called a cathedral of ironwork. It was constructed between 1859 and 1865 at the eastern end of the Southern Outfall Sewer and the Ridgeway Path, Borough of Bexley. Charles Henry Driver was its architect. Today it is a museum.

Plaque of William Lindley and Sir William Heerlein Lindley, father and son, 74 Shooters Hill Road, SE3 (Spudgun67)

William Lindley (1808–1900) and Sir William Heerlein Lindley (1853–1917) were civil engineers. The father, as a young engineer, apprenticed with Marc Isambard Brunel. He was also influenced by the teachings of the public health reformer Edwin Chadwick who advocated for building underground sewers to fight epidemics. William's three sons, among them Sir William, joined him in his efforts. They built drainage systems and underground sewers in about 30 European cities, among them Hamburg, Frankfurt (and others in German cities), St Petersburg, Budapest, Prague, Moscow, and Warsaw.

Statues of Sir Hugh Myddelton at the Holborn Viaduct, Atlantic House, north-east pavilion, EC1 (left), and by Samuel Joseph, 1845, at the Royal Exchange, on its Threadneedle Street side, EC2

Statue of Sir Hugh Myddelton by John Thomas, 1862, at Islington Green, N1, and line engraving, 1792, after C. Johnson, 1632 (Wellcome Collection)

Sir Hugh Myddelton (1560–1631) had a variety of professions but was such a skilled goldsmith that James I appointed him to be Royal Jeweler. He became a wealthy merchant and an MP. He owned mines and was a self-made engineer. To crush the ores from his lead and silver mines, he needed stamp mills, and to power them he had to build aqueducts. All this would not suffice for us to include Myddelton's memorials in this book. However, Myddelton strongly supported the development of the New River for bringing clean water from the River Lea in Hertfordshire to London. His goal was to replace the heavily polluted water from the River Thames, thereby saving lives. So dedicated was he to this project that he backed it with his own money when it encountered financial difficulties. Importantly, he also secured the King's support. The construction of the New River of 38 miles took 5 years and was finished in 1613. This is why the Myddelton memorials figure here and close to the memorials of Sir Joseph Bazalgette and the Lindleys although chronologically this would not be their proper place.

Civil Engineering

Statue of James Henry Greathead by James Butler, 1994, next to the Royal Exchange, 14 Cornhill, EC3, and a relief on the north face of the plinth showing eight men working inside the tunneling shield, and his plaque, 3 St Mary's Grove, SW13 (Spudgun67)

James Henry Greathead (1844–1896) was first educated in South Africa, then moved to England in 1859 for further education at the Westbourne Collegiate School in London. As a civil engineer, he started his career in 1869 and was involved in the design of the Tower Subway—a tunnel beneath the River Thames. His former mentor, now his superior, Peter W. Barlow (1809–1885), had the idea of a circular tunneling shield for driving tunnels, a device Greathead designed and eventually improved. The widespread use of tunneling shields was the origin of the popular name "the Tube" for the London Underground. Later, Greathead worked for the Railways where he had numerous innovations and patents for improving tunneling and other technologies. His statue stands on a high pedestal outside the Royal Exchange, above the Bank underground station, and in the middle of the road. The Underground needed a vent shaft of fairly large size for the safety regulations, and the large plinth doubles as the ventilation shaft for the Bank station below. The choice of Greathead for the statue was justified by his design of the tunneling shield technology.

Plaque of William T. Clark with his portrait on the backdrop of the Budapest Chain Bridge, Thames Path, Fulham Reach, W6 (left), and the Budapest Chain Bridge (right)

The introduction of the civil engineer William Tierney Clark (1783–1852) is a good transition from civil engineering to transportation. He designed the first suspension bridge over the Thames at Hammersmith in 1827. He also designed the first suspension bridge over the Danube connecting the two cities of Buda and Pest of what is today Budapest.[7] His designs proved durable: the Hammersmith Bridge was reconstructed by Bazalgette and was re-opened in 1887, 60 years after the original. William Tierney Clark's memorial panel depicts his portrait on the backdrop of the Budapest Chain Bridge.

Plaque of Sir John Wolfe Barry on the façade of Delahay House, 15 Chelsea Embankment, SW3 (Spudgun67), and the Tower Bridge

Sir John Wolfe Barry (1836–1918), civil engineer, designed bridges over the Thames in London. The most famous among them was the Tower Bridge, originally proposed by the architect Horace Jones (1819–1887), but completed by Wolfe Barry in 1894. Wolfe Barry is also remembered for his advocacy of and contribution to industry standardization. The mechanical engineers, naval architects, iron and steel engineers, and electrical engineers joined forces in creating what is today the British Standards Institution. A memorial window in the nave of Westminster Abbey honors him.

Sir Benjamin Baker's plaque, 3 Kensington Gate, W8

---

[7]The design was prepared in 1839 and the bridge was inaugurated in 1849. The construction was supervised by another Briton, Adam Clark (no relation). The Széchenyi Chain Bridge was blown up by the retreating Germans on January 18, 1945, along with all other bridges in Budapest, but was rebuilt in 1949, on the centenary of the original structure.

Sir Benjamin Baker (1840–1907) was a civil engineer. His best-known work was the Forth Road Bridge, about ten miles west of the center of Edinburgh. It was a joint project with another civil engineer, Sir John Fowler (1817–1898). It is a so-called cantilever bridge, the cantilever structures being beams or plates of strong and rigid material anchored at one end to a support. In 2016, it was voted in Scotland to be the greatest man-made wonder. It is a UNESCO World Heritage Site.

General view of the Vauxhall Bridge and the statue "Engineering" on one of its piers

The Vauxhall Bridge across the Thames was opened in 1906, and large statues were added to it in 1907. Frederick Pomeroy created *Agriculture*, *Architecture*, *Engineering*, and *Pottery* on the upstream piers, and Alfred Drury did *Science*, *Fine Arts*, *Local Government*, and *Education* on the downstream piers. Only one of the statues is shown here with a general view of the Bridge.

Steam Engine

Bas-relief of James Watt by John Birnie Philip/Henry Hugh Armstead on the façade of the Foreign and Commonwealth Office, and Watt's work bench from around 1800, as displayed in the Science Museum. Two of the three Watt busts are by Peter Turnerelli and the one in the middle is by Sir Francis Chantrey

James Watt (1736–1819) was a mechanical engineer and inventor who was also a skilled chemist. He is best known for his steam engine. It was an improved version of the steam engine built by Thomas Newcomen (1664–1729) in 1712. Watt received his schooling in Scotland. At the age of 18, he spent a year in London, learning instrument making. He started his career in Glasgow and became involved with repairing and eventually constructing instruments for professors of the University of Glasgow. The physicist and chemist Joseph Black and the economist Adam Smith[8] became his friends. In 1763, Watt was asked to repair the Newcomen steam engine. By this time, he was well versed in the study of heat and its motion (a science now called thermodynamics). Once he repaired Newcomen's engine and thus restored it to its original function, he made an important observation. The engine was supposed to turn heat, that is, thermal energy, into mechanical energy, but much of the thermal energy was wasted due to the construction peculiarities of the engine. Watt introduced changes, and due to his innovations, the waste heat was considerably reduced and the useful output of mechanical energy increased substantially.[9] Still, much work separated Watt's original idea and his first working model and the production of marketable engines. It took a great deal of resources and over a decade of hard work to usher in the age of steam. But Watt was dedicated and even after he met commercial success he kept improving his design for years. There were also difficulties with having his and his partner's patents recognized and collecting the fees due to them.

Watt had broad interests in making improvements in construction and in mechanical processes in a variety of fields. Thus, for example, he devised a new technique for making copies of documents. His interest in chemistry was encouraged when on a visit to Paris he learned about some chemical reactions from the famous French chemist, Claude Louis Berthollet (1748–1822). It was characteristic of Watt that, while he was trying to utilize what he had learned from others, he was also seeking ways to improve what others had invented. In his later years, he worked on techniques of copying sculptures. Among those he copied and multiplied were busts of ancient Greek scientists and philosophers as well as busts of himself.

James Watt was a remarkable contributor to the Scientific Revolution. The inscription on the cenotaph does not exaggerate when it proclaims, "James Watt … enlarged the resources of his country, increased the power of man, and rose to an eminent place among the most illustrious followers of science and the real benefactors of the world."[10]

---

[8] Adam Smith (1723–1790) was an economist of great renown; He has a statue at Burlington Gardens (Chap. 1).

[9] We note that the considerable reduction of waste did not mean full elimination of waste; even today absolute conversion is not achieved. This waste heat is an important contributor to the climate change.

[10] https://upload.wikimedia.org/wikipedia/commons/5/50/Watt_James_Chantrey.jpg (downloaded July 31, 2019).

Transportation

By Rail

Plaque of Richard Trevithick, Gower Street, WC1E, and his bust at the Science Museum

Richard Trevithick (1771–1833) was an engineer and inventor who came from a family of miners. Already in his childhood he was fascinated with steam engines. As a young adult, he worked in the mining industry and built engines sufficiently different from Watt's to avoid patent fees. Trevithick was among the first experimenting with high pressure steam and introduced technical innovations. At some point, he attached his high pressure steam engine to a road carriage, and by 1801 he had built a full-size locomotive, which he called "Puffing Devil." He kept improving his design, and, following an explosion, augmented it with safety features. In 1808, he demonstrated his latest locomotive in London on a circular track; spectators were charged a fee to see the spectacle. By 1812, Trevithick's locomotives started replacing horses, first in moving coal wagons, but eventually, passenger carriages. Trevithick was involved in the initial attempts of building a tunnel under the Thames and other projects. His memorials stand in many places, including his full-size statue in front of the Passmore Edwards Free Library in Camborne, Cromwell. A replica of his first locomotive was displayed in 2001 in Camborne, where he first demonstrated it 200 years before.

Statue of Robert Stephenson by Carlo (Charles) Marochetti, 1871, in front of Euston Station, WC1, and his engraving by D. J. Pound, 1860, after J. E. Mayall (Wellcome Collection)

Robert Stephenson (1803–1859) was a civil engineer who built bridges and railways. He has been called the greatest engineer of the nineteenth century. His father, George Stephenson (1781–1848), built the first public railway line to use steam locomotives and has been called the Father of Railways. Robert especially excelled in building bridges, both in Britain and in other lands. He presided over a great expansion of railway building: at one time he was in charge of one-third of all railway construction in Britain. The London-Birmingham Railway was one of his projects jointly with his father. Euston Station is the London terminus of this line and Robert's statue stands in its front. There is a scroll in his right hand, representing bridge or railway designs. There used to be also a George Stephenson statue in the main entrance of Euston Station, but it was moved to the National Railway Museum in York. George Stephenson has a portrait at the National Portrait Gallery, painted by Henry William Pickersgill. Robert Stephenson is buried in the nave of Westminster Abbey.

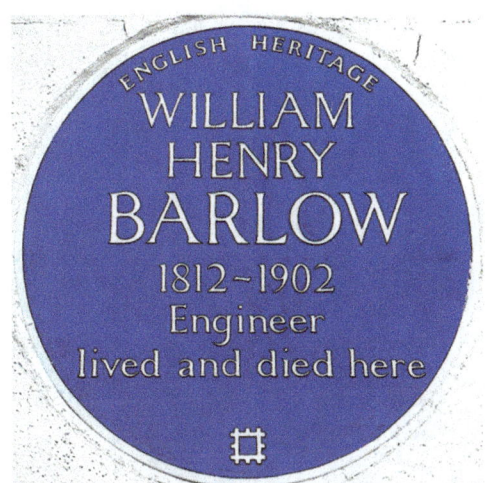  

Left: Plaque of William Henry Barlow, 145 Charlton Road, SE7 (Spudgun67). Middle and right: Plaque of Sir Nigel Gresley (Spudgun67) and his statue by Hazel Reeves (Spudgun67) at Platform 8 of King's Cross Station

William Henry Barlow (1812–1902) was a civil engineer engaged mainly in railway engineering. He invented new rail designs, liked to experiment, and investigated steel structures.

Sir Nigel Gresley (1876–1941) was a locomotive engineer who had a brilliant career in the British railway system. Not only did he invent a new three-cylinder design, but his locomotives were also efficient mechanically and elegant in appearance and became famous after setting many speed records. His "Flying Scotsman" was the first locomotive whose speed exceeded 100 mph in passenger service, and his "Mallard" has kept its world record, being the fastest ever steam locomotive. The name "Mallard" hinted at his hobby of breeding fowl; in fact, he named several of his locomotives after birds. His statue at King's Cross originally displayed a duck, but in the final execution, the bird was not realized. The sculptor removed it after Gresley's grandsons protested, considering it undignified.

By Water

Plaque of Sir Francis Pettit Smith, 17 Sydenham Hill, SE26 (Spudgun67)

Sir Francis Pettit Smith (1808–1874) was a farmer most of his life, but he was also an inventor. He invented the screw propeller, which converts rotational motion into thrust. Its concept has a long history; many others made similar inventions, independently. The Swedish-American John Ericsson (1803–1889) is perhaps the best known among them. Pettit Smith built models of and patented his screw propeller. He was instrumental in bringing to completion the *SS Archimedes,* the first steamship equipped with his invention.

By Road

Plaques of Charles Rolls, 14/15 Conduit Street, W1S (left), and Sir Harry Ricardo, 13 Bedford Square, WC1 (right)

Charles Rolls (1877–1910) was not an engineer, but his activities greatly promoted motoring and later aviation in the United Kingdom and beyond. He held many records in motoring and flying. He was the first man in England who owned a car. He was a founding member of the predecessor of what is today the Royal Automobile Club. He met the engineer Henry Royce in 1904, and 2 years later they formed the Rolls-Royce Company. Rolls was also a founding member of the Royal Aero Club and was one of the first pilots who flew a newly invented plane of the Wright Brothers. While not yet 40, he was killed in an accident during a flying tournament.

Sir Harry Ricardo (1885–1974) was an engineer best known for his work designing internal combustion engines. He started building engines as a teenager. At 18, he enrolled at Trinity College in Cambridge. Beside motorcycle and car engines, he developed tank and aircraft engines. His innovations helped the British Military, but monitoring his patents, Germany also made good use of them prior to World War II. During the war, Ricardo was a member of the engineering advisory committee of the War Cabinet. Even in retirement, after 1964, he remained active in consulting and advising.

By Air

 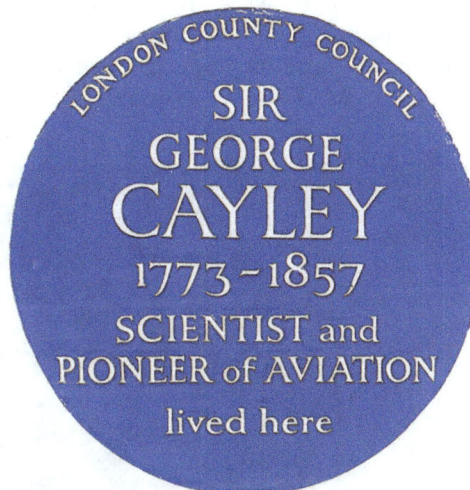

Left: Plaque commemorating the founding of the Royal Polytechnic Institution by George Cayley in 1838–1839, 309 Regent Street, W1. Right: Plaque of Sir George Cayley, 20 Hertford Street, W1

Sir George Cayley (1773–1857) was an engineer and a pioneer of aviation, who helped place aviation on a scientific foundation. Many of the basic concepts of today's aviation can be traced back to his studies. He was concerned with the education of engineers and established the Royal Polytechnic Institution in 1838–1839, the first such institution of higher education in the United Kingdom. He was the first chairman of the School. Between 1881 and 1992, it operated under the name of Polytechnic at Regent Street or Polytechnic of Central London and, since then, has become known as the University of Westminster of which Regent Street is one of the campuses. Sir Alexander Fleming was among its alumni.

From left to right: Plaque commemorating the expansion of the Royal Polytechnic Institution by Quintin Hogg in 1881–1882, 309 Regent Street, W1; memorial for Quintin Hogg, Alice Hogg, and the fallen heroes of the Polytechnic in the wars 1914–1918 and 1939–1945, by George Frampton, 1906; and plaque of Quintin Hogg, 5 Cavendish Square, W1

The Eton-graduate and wealthy Christian educational reformer Quintin Hogg (1845–1903) played a pivotal role in educational reform. The Regent Street Polytechnic he helped found incorporated the Royal Polytechnic Institution and is now the University of Westminster. His memorial depicts him reading to two children. It stands opposite the BBC Broadcasting House and honors not only him but also his wife, Alice, and the fallen members of the Polytechnic in World Wars I and II.

Two plaques of A. V. Roe. Left: At the intersection of West Hill and Cromford Road, SW15 (Matt Brown). Right: At the railway arches of the Walthamstow Marsh Railway Viaduct, E17 (Spudgun67)

Sir Edwin Alliott Verdon Roe (1877–1958) was a test pilot, aircraft designer, and aircraft manufacturer. His attempt for admission to study marine engineering at King's College failed. But he was an early enthusiast of flying, and at some point Charles Ross helped him with his career. Verdon Roe (sometimes, Verdon-Roe) experimented with flying models and built an airplane, which he flew successfully in 1908. He co-founded an aircraft company which sold thousands of planes to the Royal Flying Corps, later, Royal Air Force (RAF) for training purposes. Verdon Roe was a member of the British Union of Fascists (BUF) and a supporter of Sir Oswald Mosley, the leader of BUF. Notwithstanding the questionable politics of their father, two of his sons, both squadron leaders of the RAF, were killed in action during WWII.

Plaques of Sir Frederick Handley Page, 18 Grosvenor Square, W1K (left, Matt Brown), and the three Short brothers at Arch 75, Queen's Circus, SW8 (right, Spudgun67)

Sir Frederick Handley Page (1885–1962) designed and manufactured aircraft, especially heavy bombers. His best-known military planes were the Handley Page 0/400 bomber, deployed in WWI, and the Halifax bomber, deployed in WWII. Between the two world wars, his plane, the H.P.42, was the flagship of the Imperial Airways, one of the companies whose mergers eventually formed British Airways in 1974. Remarkably, the H.P.42 was never involved in any passenger deaths, a testament to the foresight and caution of its engineer. Handley Page and the German Gustav Lachmann independently invented an important improvement of aircraft wing. Rather than getting into legal battle for patent priorities, the two inventors agreed to a shared ownership and an amicable cooperation resulted between the two in the company owned by Handley Page.

The three Short brothers, Horace (1872–1917), Eustace (1875–1932), and Oswald (1883–1969), were aeronautical engineers. They helped Charles Rolls build his balloon for an international race. They also experimented with hot air balloons, but soon switched to airplanes. They established their workshop at the railway arches of Queen's Circus in Battersea Park and founded the world's first company for producing airplanes, manufacturing military aircraft for the British efforts both in WWI and WWII. Their principal manufacturing base was in Northern Ireland. They were influential in their efficient training of engineers and engineering technicians in cooperation with Irish institutions of higher education.

Plaques of Sir Geoffrey de Havilland, 32 Baron's Court Road, W14 (left, Spudgun67), and Sir Thomas Sopwith, 46 Green Street, W1K (right, Spudgun67)

Sir Geoffrey de Havilland (1882–1965) was an aviator and aerospace engineer. He started building airplanes in the pioneering time of flying. His aircraft were used both in WWI and WWII. He designed the most versatile warplane, the de Havilland DH.98 Mosquito, and the world's first commercial jet airliner, the de Havilland DH.106 Comet. There is a de Havilland Heron plane on display as a monument to the role of Croydon Airport in aviation history. It stands in front of Airport House, Purley Way, South Croydon, CR0. The Airport House used to be the terminal building of Croydon Airport, which had a pioneering venue in civilian aviation. At one time it was London's main airport. In 1930, Amy Johnson started her historic solo flight to Australia from this airport. In WWII, Croydon Airport served exclusively as a Royal Air Force fighter station.

Sir Thomas Sopwith (1888–1989) first experienced flying in a hot air balloon of C. S. Rolls. Then, Sopwith bought his own balloon from the company of the Short brothers. He learned to pilot airplanes, founded his own manufacturing company, introduced innovations in airplane design, and established flight records.

Amy Johnson' plaque, Vernon Court, Hendon Way, NW2 (Spudgun67), and her portrait (Wikimedia)

Amy Johnson (1903–1941) was called the "Queen of the Air." She studied at Sheffield University and became interested in aviation when, following graduation, she moved to London in 1927. She soon became a licensed pilot and started establishing records. She flew solo from England to Australia in 1930 as the first woman accomplishing such a deed. Other record-breaking flights followed during the 1930s. In WWII she served as a pilot of the Air Transport Auxiliary. She disappeared over the Thames under poor weather conditions. Her body was never recovered.

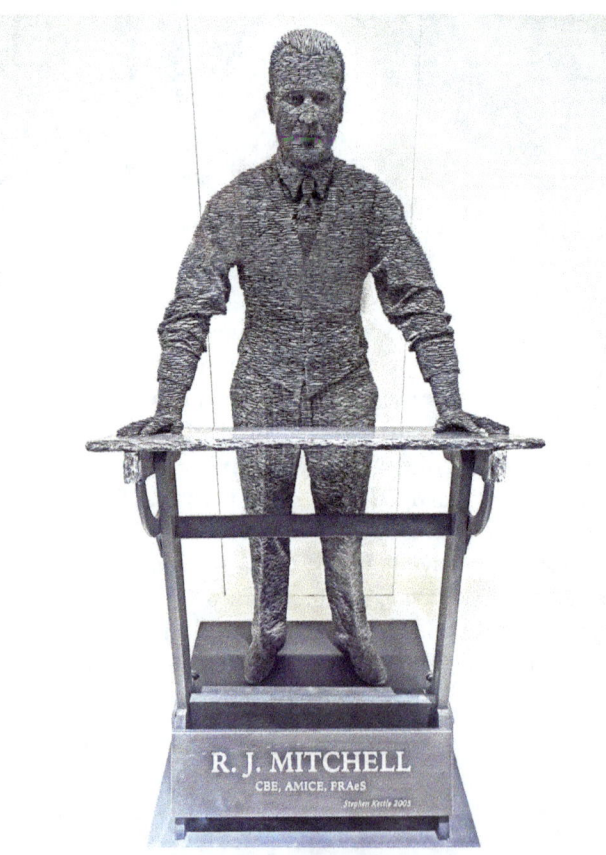

Reginald Joseph Mitchell's sculpture by Stephen Kettle, 2005, at the Science Museum

Reginald Joseph Mitchell (1895–1937) was an engineer and aircraft designer. As the Chief Designer of the Supermarine aircraft company in Southampton, he designed 24 different aircraft. The most successful and best known was the Spitfire, much deployed during the *Blitz* by the Royal Air Force. Mitchell's early death from cancer was a great loss to Britain's war effort.

Plaques of Sir Robert Watson-Watt, 287 Sheen Lane, SW14, and Alan Dower Blumlein, 37 The Ridings, W5 (both, Spudgun67)

Sir Robert Watson-Watt (1892–1973) studied at what is today the University of Dundee. While earning prizes in chemistry and in natural philosophy, he graduated as an engineer. Radio physics was his primary interest. He first applied radio waves in meteorology, tracking approaching thunderstorms. Through a series of technical innovations, he and his junior colleague, Arnold Frederic Wilkins (1907–1985), recognized the possibility of detecting aircraft with radio waves, even at very large distances. They proved the feasibility of radar technology as early as the mid-1930s. In 1940, their secret weapon of radar helped the Royal Air Force win the Battle of Britain. The invention, which was further perfected, played a significant role in the eventual Allied victory in WWII. In addition to British recognition, Watson-Watt was awarded the US Medal of Merit in 1946.

Alan Dower Blumlein (1903–1942) was an electronics engineer and a war hero. He was already tinkering with various devices from the age of 7. At Imperial College he worked in telecommunications, sound recording, television, and radar. In each of these areas and elsewhere he had innovations. In WWII, he was working on improving radar when his plane crashed during a test flight and he was killed. In addition to the plaque shown above, there is a rectangular tablet commemorating Blumlein's invention of stereo sound reproduction, the two-channel audio system, at 3 Abbey Road, NW8.

Clockmakers

Beside the importance of clock making in everyday life, it had outstanding significance for navigation and astronomy—one reason for the great respect with which heads of state treated their clockmakers. They are remembered with plaques and busts and several of them were buried in Westminster Abbey.

Portrait of Nicholas Kratzer (line engraving by F. J. Dequevauviller after G. Anastasi after H. Holbein, Wellcome Collection)

King Henry VIII (1491–1547) was keen to have the best clocks in his court. At least two outstanding clockmakers were in his employ. There is an astronomical clock at the Anne Boleyn Gate, Clock Court, Hampton Court Palace. It was designed by Nicholas Kratzer (1487–c.1550) and made by a Huguenot immigrant, Nicholas Oursian. The clock depicts the sun revolving around the earth. It went into operation in 1540, just a few years before Copernicus published his cosmological theories. Oursian was a disciple of Kratzer and served as the Keeper of Clocks at Hampton Court. The designer Kratzer was a Bavarian mathematician whom Henry VIII invited to his court. He was also a lecturer at Oxford and tutor to Sir Thomas More's children. It is supposed that the great sundial at the palace of Whitehall was Kratzer's design. The astronomical clock at Hampton Court was restored several times during its long existence; the latest was in 2008 in preparation for the 2009 celebrations of the 500th anniversary of Henry VIII's accession to the throne.[11]

Some outstanding clockmakers are mentioned here and in the section about Greenwich in Chap. 2. There are 8 modest plaques on the façade of 33 St John's Lane, EC1, called also "Watchmakers' Court." Facing the building the plaques are arranged chronologically, from left to right: Thomas Tompion (1639–1713). Christopher Pinchbeck (1670–1732). Joseph Simms (1745–1770). James Upjohn (1765–1794). John Cranfield (1770–1790). Edward Massey (1772–1852). John Moor (1801–1875), and Dan Parkes (1946–1989).

---

[11] Frank Lloyd and Helen Potkin, Davina Thackara, *Public Sculpture of Outer South and West London* (Public Sculpture of Britain Volume Thirteen, Liverpool: Liverpool University Press, 2011), pp. 212–213.

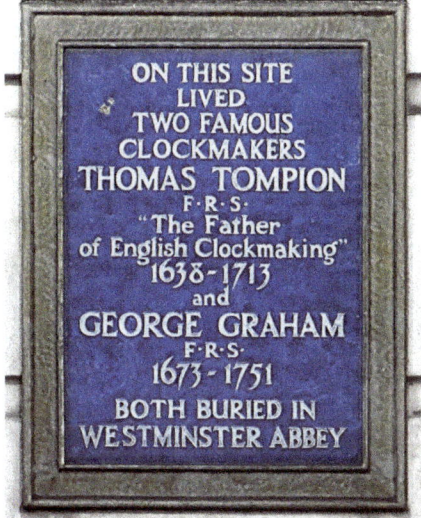  

From left to right: Tablet of Thomas Tompion and George Graham, 69 Fleet Street, EC4 (left); statue of Thomas Tompion in a niche on the façade of the Victoria and Albert Museum (Spudgun67); and a portrait of George Graham, by J. Tookey (Wellcome Collection)

Plaques of John Harrison, Summit House, Red Lion Square (Dane Street), WC1R, and Thomas Earnshaw, 119 High Holborn, WC1

John Harrison (1693–1776) was to follow his father's profession and become a builder. However, he developed an interest in cleaning and repairing watches. He even made some new ones and introduced innovations in watch design. In 2006, a memorial stone was unveiled honoring "John 'Longitude' Harrison, Clockmaker 1693–1776." There is a bimetallic strip running through the memorial stone at the longitude line 000 07′35″ W, the one that ran through Harrison's life. The meaning of this and the 'Longitude' as his "middle name" has a story. The determination of longitude at sea used to be a major difficulty. The Longitude Act of 1714 by the Parliament of the United Kingdom offered a reward of £20,000 for a reliable method to determine it. Harrison heard about it in 1726 and he worked on the problem for 30 years. Finally, he solved it, developing a reliable marine chronometer.

Thomas Earnshaw (1749–1829) and another watchmaker, John Arnold (1736–1799), were also pioneers of the development of reliable chronometers. They continued the work of John Harrison and designed watches (pocket chronometers; Arnold brought the term "watch" into its modern usage) that were both accurate and practical. Each of the two watchmakers was awarded a prize in 1805 for the improvements of their chronometers. For Arnold, this was a posthumous recognition. The prize-giving institution was the Board of Longitude, established by the same Act of Parliament that, in 1714, offered the prize for solving the problem of the determination of longitude at sea.

Among the many exhibits of clockmakers of earlier times in the Science Museum, there are paintings and a bust of a modern one, George Daniels (1926–2011). For describing his profession, the terms watchmaker and horologist are used. He improved the design of watches by using a new type of escapement—mechanical linkage—making lubrication superfluous. His "Space Traveller's Watch I" has been so far the most expensive watch, sold in 2019 in London for £3,615,000.

Bust of George Daniels by Sir Eduardo Paolozzi, 1997, at the Science Museum

Other Inventors

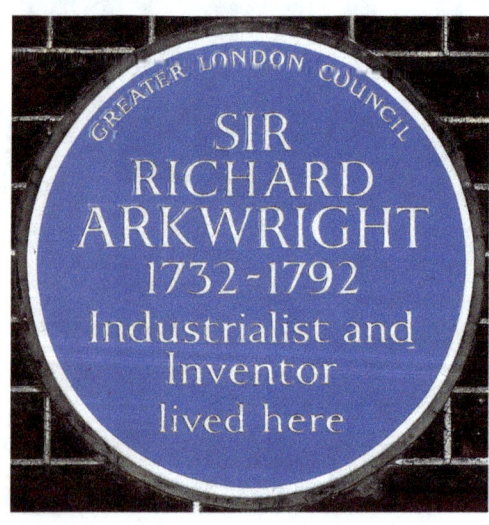

Plaques of Sir Richard Arkwright, 8 Adam Street, WC2N, and Charles 3rd Earl Stanhope, 20 Mansfield Street, W1G (Matt Brown)

Sir Richard Arkwright (1732–1792) was an industrialist and inventor who improved and patented a rotary carding engine for disentangling cotton fibers and for producing a continuous thread for subsequent processing. Furthermore, he contributed to the mechanization of the entire manufacturing process of cotton production by introducing a number of innovations. Because he was one of several contributors to the mechanization of the cotton industry, he was subjected to lengthy court procedures as patent rights were litigated. As one of the initiators of the concept of modern factory, he has been called "the father of the modern factory." His activities constituted an important component of the fledgling Industrial Revolution.

Charles Stanhope, 3rd Earl Stanhope (1753–1816), attended Eton College, then studied mathematics and electricity at the University of Geneva. Upon his return to England, he devoted his considerable wealth to experimental studies and to developing inventions. These included a technique of fire protection for wooden buildings (this was not adopted), the iron printing press, a lens bearing his name, and a device for tuning musical instruments. He attempted constructing calculating machines. He was also a politician. His first marriage was to the daughter of the Prime Minister William Pitt ("Pitt, the Elder"). One of their daughters, Lady Hester Lucy Stanhope, became a renowned traveler and scholar in Arabic studies. His brother-in-law, William Pitt the Younger, was another Prime Minister.

For the renowned inventor, Sir Benjamin Thompson—Count Rumford, see Chap. 3.

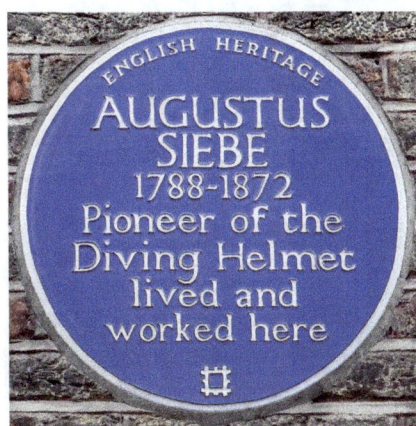  

From left to right: Plaques of Augustus Siebe, 5 Denmark Street, WC2H (Spudgun67); Charles Manby, 60 Westbourne Terrace, W2 (Spudgun67); and Edward Goodrich Acheson, 31 Prince Albert Road, NW8 (Matt Brown)

Augustus Siebe (1788–1872) was born in Germany and graduated in Berlin as an engineer. He served in the Prussian Army in the Napoleonic wars. Afterward he moved to London and became an instrument maker. Of his many innovations, the best known is his diving gear, which included a diving helmet, utilized for underwater work.

Charles Manby (1804–1884) learned engineering with iron, and in 1823, he started installing gas pipes in Paris and elsewhere. He was involved in a host of other civil engineering assignments, founded companies, and developed ventilation systems for large buildings. He interacted with Samuel Colt in the manufacture of firearms and with Robert Stephenson whose rail enterprises Manby represented throughout Europe. He was active in international institutions thanks to his engineering skills and his perfect knowledge of the French language.

Edward Goodrich Acheson (1856–1931) was an American inventor whose patents included an electric furnace. He was a self-educated scientist who left school early on account of helping his impoverished family. But he was a natural experimenter who managed to sell a battery to Edison, for whom he worked briefly. Acheson lived in London only for a few years at the time of WWI, but in later travels in Europe, he established the lighting systems of Antwerp, Belgium, and in the opera house La Scala in Milan. He produced silicon carbide, called carborundum, using the so-called Acheson procedure he invented.

Plaque of Frederick Winsor, 100 Pall Mall, SW1 (left), and one commemorating the Gas Light & Coke Company at the junction of Great Peter Street and St Anne's Street, SW1 (this image, Matt Brown)

Frederick Winsor (Friedrich Winzer, 1763–1830) was a German inventor who was keenly interested in British and French progress in the technology of fuels. He spent some time in London and Paris. In London, he was researching the possibilities of using coal gas for illumination, and in 1807 he lit one side of Pall Mall with such gas lamps. This event is commemorated with the memorial plaque. Another green plaque commemorates the Gas Light & Coke Company, a gas company that between 1813 and 1837 provided a public supply of gas for the first time anywhere in the world. It later became the British Gas plc.

Statue of Sir Corbet Woodall by George Arthur Walker in the garden at Twelvetrees Crescent, Bromley by Bow, E3 (Matt Brown)

Sir Corbet Woodall (1841–1916) was a gas engineer who helped devise schemes for hydraulic power networks in the United Kingdom and Australia. He worked as company director and as private consultant at various stages of his career. He was the Governor of the Gas Light & Coke Company during the last decade of his life.

Sir Hiram Maxim's plaque, Hatton Garden, EC1, on the wall of Kovacs House, close to the corner of Hatton Garden and Clerkenwell Road (Matt Brown)

Sir Hiram Maxim (1840–1916) was an American-born British engineer who invented the Maxim gun in 1884. It was the first portable, fully automated machine gun and was capable of firing up to 600 rounds per minute. Maxim had numerous, more innocuous inventions, among them a hair-curling iron, a mousetrap, and a steam pump. A number of his patents described electric lamps.

Stained window honoring Sir Henry Tizard at the RAF Bentley Priory Museum, near Stanmore, in the Borough of Harrow (Matt Brown)

Sir Henry Thomas Tizard (1885–1959), engineer and chemist, studied at Westminster School and at Oxford. He completed his science education in Berlin, where he formed a friendship with Frederick Alexander Lindemann, Winston Churchill's future science advisor during WWII. Upon returning to England, Tizard worked at the Davy-Faraday Laboratory of the Royal Institution, then, at Oxford. Working for a while for the Shell Company, he invented the concept of what is called today the octane number for rating the performance of fuels in internal combustion engines. He was a long-time Rector of Imperial College and held various science-related positions in the Department of Defence. He helped with the development of radar. In wartime, he led the so-called Tizard Mission to the United States. Its task was to introduce the British technological advances to the Americans in the framework of the extensive US-British cooperation in science and technology during WWII. The resonant-cavity magnetron, the heart of radar, the Whittle gas turbine, and the British achievements in nuclear research were among those British advancements shared. Throughout the years of WWII and during the post-war years, Tizard held leading positions in the defense-related science establishment. When various reports referred to the appearance of UFOs, it was Tizard's decision that a serious investigation should precede the dismissal of those reports.

Dennis Gabor's plaque, 79 Queen's Gate, SW7, and portrait (courtesy of the late George Marx)

Dennis Gabor (1900–1979) was a Hungarian-born British engineer-physicist and a Nobel laureate in Physics for the invention of holography. He was interested in innovation from his childhood. When he was just 10 years old, he filed for a patent about his invention in connection with a carousel (merry-go-round), which was granted in 1911. He graduated from an excellent high school in Budapest, but spent only a few months at the Faculty of Mechanical Engineering of the Budapest Technical University. In 1920, at the dawn of the autocratic and anti-Semitic Horthy regime, he moved to Germany and completed his studies at the Technical University of Berlin. He attended the famous physics colloquia and Albert Einstein's seminars at the University of Berlin. He was a member of the informal circle of Hungarian émigré intellectuals that also included Leo Szilard, John von Neumann, Eugene P. Wigner, Michael Polanyi, and Arthur Koestler.

After the Nazi takeover of Germany, the Jewish Gabor moved to England, where he had research and development jobs in industrial laboratories. It was while working in industry that he came to the idea of holography, which he then developed and communicated between 1947 and 1951. From 1948, he secured an appointment at Imperial College, where he rose to Professor of Applied Physics in 1958. He retired in 1967. In retirement he continued his research in connection with Imperial College and the CBS Laboratories in Stamford, Connecticut, especially in technologies of communication and display. He became also interested in social issues and authored books.

"The Head of Invention" by Sir Eduardo Paolozzi, 1989, in the garden of the Design Museum, 224–238 Kensington High Street, W8

Although the original inspiration for Paolozzi's "The Head of Invention" came from a study of James Watt's head, the sculptor's message is in the words attributed to Leonardo and carved into the side of the sculpture: "Though human genius in its various inventions with various instruments may answer the same end, it will never find an invention more beautiful than Nature, because in her inventions nothing is lacking and nothing is superfluous."

Bibliography

Barnett, Richard, *Anatomy of the City: A Guide to Medical London* (London: Strange Attractor Press, 2008a)

Barnett, Richard, *Sick City: Two thousand years of life and death in London* (London: Strange Attractor Press, 2008b)

Black, Nick, *Walking London's Medical History* (London: Royal Society of Medicine Press, 2006)

Cavanaugh, Terry, *Public Sculpture of South London* (Public Sculpture of Britain Volume Ten, Liverpool: Liverpool University Press, 2007)

Hall, A. Rupert, *The Abbey Scientists: The Memorials of Westminster Abbey* (London: Roger and Robert Nicholson, 1966)

Heart of a London Village: The Highgate Literary and Scientific Institution 1839–1990 (The Highgate Literary and Scientific Institution, 1991)

Jackson, Roland, *The Ascent of John Tyndall: Victorian Scientist, Mountaineer, and Public Intellectual* (Oxford: Oxford University Press, 2018)

Johnson, Steven, *The Ghost Map: The Story of London's Most Terrifying Epidemic—and How It Changed Science, Cities, and the Modern World* (New York: Riverhead Books, 2006)

Kedar, Benjamin Z., ed., *Chaim Weizmann: Scientist, Statesman, and Architect of Science Policy* (Jerusalem: The Israel Academy of Sciences and Humanities, 2017)

Kershman, Andrew, *London's Monuments* (London: Metro Publications, 2013)

Lloyd, Fran, Helen Potkin, Davina Thackara, *Public Sculpture of Outer South and West London* (Public Sculpture of Britain Volume Thirteen, Liverpool: Liverpool University Press, 2011)

Long, David, *Bizarre London: Discover the Capital's Secrets & Surprises* (New York: Skyhorse Publishing, 2013)

Marshall, Geoff, *London's Industrial Heritage* (The History Press, 2013)

Matthews, Peter, *London's Statues and Monuments* (Oxford: Bloomsbury Shire Publications, 2012)

Matthews, Peter, *Who's Buried Where in London* (Oxford: Bloomsbury Shire Publications, 2017)

McGovern, Una, ed., *Chambers Biographical Dictionary* (7th Edition, Edinburgh: Chambers Harrap Publs, 2002)

Olby, Robert, *Francis Crick: Hunter of Life's Secrets* (Cold Spring Harbor, NY: Cold Spring Harbor Laboratory Press, 2009)

Popper, Karl, *The Logic of Scientific Discovery* (London and New York: Routledge Classics, 2002; reprint of the 1959 first edition in English)

Rennison, Nick, *The London Blue Plaque Guide* (The History Press, 2009)

Ridley, Matt, *Francis Crick: Discoverer of the Genetic Code* (New York: HarperCollins, 2006)

Rosen, Dennis and Sylvia, *London Science: Museums, libraries, and places of scientific, technological & medical interest* (London: PRION, 1994)

Smith, Denis, *Civil Engineering Heritage: London and the Thames Valley*, MPG Books, Bodmin, 2001

Sykes, John, *111 Places in London that you shouldn't miss* (Emons Verlag, 2017)

Tinniswood, Adrian, *The Royal Society and the Invention of Modern Science* (New York: Basic Books, 2019)

Ward-Jackson, Philip, *Public Sculpture of Historic Westminster Volume 1* (Public Sculpture of Britain Volume Fourteen, Liverpool: Liverpool University Press, 2011)

Ward-Jackson, Philip, *Public Sculpture of the City of London* (Public Sculpture of Britain Volume Seven, Liverpool: Liverpool University Press, 2003)

Westminster Abbey: Official Guide (© 1997 Dean and Chapter of Westminster; our copy is the revised edition of 2002)

Index of Names

A
Acheson, E.G., 265
Addison, T., 42, 148
Airy, G., 47, 50, 51
Albert, P., 7–12, 243, 265
Aldrich-Blake, L., 174
Alexandra, P., 183
Alfred, P., 8
Amundsen, R., 45
Anderson, E.G., 171, 172, 174, 227
Anderson, J., 172
Anne, P., 78
Anning, M., 128, 227, 238
Appleton, E.V., 92
Arber (née Robertson), A., 137
Archimedes, 79, 94, 256
Aristotle, 3, 18
Arkwright, R., 264
Arthur, C.A., 51
Asclepius, 141, 142, 161
Avery, O., 197
Ayrton (née Marks), H., 226
Ayrton, W.E., 227

B
Babbage, C., 17, 47, 74, 230, 231
Babington, W., 147
Bacon, F., 3, 4, 6, 94, 145
Bacon, R., 94
Bader, A., 25
Baird, J.L., 228, 229
Baker, B., 251
Balfour, A., 169, 240
Balint, E., 208
Balint, M., 208
Banks, J., 15, 69, 70, 124, 125, 129, 137
Barnard, C., 59, 148
Barnes, B., 138
Barnes, R., 151
Barry, J.W., 250
Bartimaeus, 193
Barton, D., 117, 118
Barton, E.R., *see* Eustace, R.
Báthory, S., 35
Bazalgette, J.W., 241, 244, 245, 248
Beaufort, F., 43, 46, 47
Becher, J., 102
Beck, H., 239
Becket, T.à, 153
Beethoven, L. van, 75
Beit, A., 236

Bellot, J.R., 43, 44
Bentham, G., 128, 129
Bentham, J., 19
Bernal, J.D., 99, 170
Berthelot, M., 109
Berthollet, C.L., 252
Berzelius, J.J., 108
Bessemer, H., 238
Beveridge, W.H., 23
Biggs, H.M., 165
Black, James, 152, 252
Black, Joseph, 252
Blackett, P.M.S., 93
Blackie, M., 176
Blackwell, E., 171, 172, 174
Blumlein, D., 261
Bond, M., 235
Bose, J., 188
Bourbons, 85
Bowers, H.R., 45
Bowman, W., 194
Boyle, R., 58, 61, 65, 75, 101, 102
Bragg, W.H., 23, 77
Bragg, W.L., 77, 121
Brandt, H., 102
Braun, K.F., 228
Brenner, S., 10
Bright, R., 147, 148
Brill, A.A., 206
Bristowe, J.S., 159, 160
Brouncker, W., 60
Brout, R., 100
Brown, R., 125
Bruce, D., 165
Bunsen, R.W., 75, 109, 240
Burnet, M., 191
Burroughs, S.M., 216
Burton, R.F., 39
Byron, L., 171

C
Caine, M., 239
Carlyle, T., 75
Caroline, Queen, 157
Carter, H., 182
Cavell, E., 179
Cavendish, H., 70, 77, 90, 92, 104, 105
Cavendish, W., 105
Cayley, A., 134
Cayley, G., 257
Chadwick, E., 165, 186, 246

Chain, E.B., 116, 117, 199
Charles I, 7, 61, 145
Charles II, 7, 60, 64, 146
Cheselden, W., 157
Chichester, F., 33, 34
Christie, W.H.M., 134
Churchill, W., 95, 97
Cicero, G., 19
Clare, A., 152
Clark, A., 250
Clark, Fredrick Le Gros, 158
Clark, W.T., 249, 250
Clover, J., 182, 183
Cole, H., 8, 9
Columbus, C., 33, 145
Conan Doyle, A., 15
Conolly, J., 147, 148
Cook, J., 27, 34–37, 49
Cooke, W.F., 222, 223
Cooper, A.P., 163
Copeman, S.M., 190
Copernicus, N., 6, 94, 145, 262
Cornforth, J.W., 10
Cotton, R., 15
Cranfield, J., 262
Creed, F.G., 227
Crick, F., 119–121
Crompton, R.E.B., 221, 222
Cromwell, O., 16, 24, 253
Crookes, W., 109–111, 134
Cubberley, E.P., 142
Culpeper, N., 214, 215
Cuvier, G., 18, 126

D
da Vinci, L., 11, 219, 220
Dale, H., 148, 197
d'Alembert, Jean le Rond, 85
Dallos, J., 195
Dalton, J., 106
Dampier, W., 35
Daniell, J.F., 223
Daniels, G., 264
Dante, 79
Darwin, C., 22, 47, 56–138, 145, 148, 239
Darwin, E., 149
Darwin, G.H., 134
Davisson, C., 91
Davy, H., 70, 72, 110, 240, 268
de Broglie, L., 138
de Havilland, G., 259
de la Beche, H., 238
Dee, J., 6
Descartes, R., 82
Dewar, J., 76
Dias, B., 33
Dick-Read, G., 204
Dirac, P., 21
Disraeli, B., 109
Don, D., 125
Down, J.L., 202
Drake, F., 33, 34, 49
Drake, N., 234
Drysdale, V., 175, 176
Duchenne de Boulogne, G.B.A., 201
Duke-Elder, S., 194

Dürer, A., 11
Dyer, W.T.T., 134
Dyson, F., 54, 96

E
Earnshaw, T.A.J., 263
Eddington, A.S., 54, 96
Edison, T.A., 91, 226, 265
Edward VI, 58, 153, 155
Edward VII, 12, 148, 160, 183, 187
Edward, P., 183
Ehrlich, P., 199
Eijkman, C., 114
Einstein, A., 24, 82, 90, 95, 96, 125, 145
Elizabeth I, 5, 7, 34, 58
Elizabeth II, 78, 176, 194
Elliotson, J., 151
Ellis, H.H., 207
Englert, F., 100
Ericsson, J., 256
Ernster, L., 26
Euclid, 74
Eustace, R., 119
Evans, E., 45
Evans, J., 134

F
Faraday, M., 8, 70, 71, 73, 74, 76, 111, 202
Farnsworth, P.T., 229
Farr, W., 165
Fawcett, H., 132
Fenwick, E.G., 179
Ferenczi, S., 206, 208
Fischer, E.O., 117
Fischer, R.A., 90
Fitzroy, R., 37, 46, 47, 130
Flamsteed, J., 50, 51
Flamsteed, M., 51
Flaxman, J., 19, 220
Fleming, Alexander, 148, 198, 199, 257
Fleming, Ambrose, 226
Flinders, M., 37, 125
Florey, H.W., 116, 199
Flower, W.H., 134
Forssmann, W., 145
Fortune, R., 132, 133
Fowler, J., 251
Frank, J.P., 166
Frankland, E., 109, 134
Franklin, B., 104, 221
Franklin, J., 41–43, 47, 48
Franklin, R.E., 120, 121, 170, 227
Fredericq, P., 116
Freud, A., 206
Freud, S., 148, 205, 206
Friedrich III, Emperor, 191
Fry, J., 162
Fry, K., 152
Funk, C., 114

G
Gabor, D., 268
Galen, *see* Galenus, Aleius/Claudius
Galenus, Aleius/Claudius, 19, 208

Index of Names

Galilei, G., 18, 80, 94, 145
Galton, F., 90, 134, 143, 144
Gates, F.T., 149
Gay-Lussac, J.-L., 108
Geber, 94
Geikie, A., 134
George II, 14, 15
George III, 7, 15, 125, 209
George V, 192
George VI, 183, 192
George, L., 95
Germer, L., 91
Gestetner, D., 227
Gilbert, W., 6, 94
Gillespie, R.J., 115
Gillies, H., 213, 214
Gladstone, W., 109, 238
Glaisher, J., 53
Godfrey, A., 102
Godman, F.D., 136
Goethe, J.W. von, 11, 18, 86, 137
Gorgas, W.C., 166
Gosling, R.G., 120, 121
Gosse, P.H., 132, 133
Graham, G., 263
Graham, T., 101
Grant, R.E., 127
Grassi, G.B., 169
Gratia, A., 116
Gray, H., 3, 180, 181
Greathead, J.H., 249
Green, G., 21
Grenville, T., 15
Gresham, T., 58, 59, 64, 65, 90
Gresley, N., 255
Gutenberg, J., 145
Guy, T., 160, 161, 163
Gwynne-Vaugham (née Fraser), H., 137, 138

H
Haldane, J.B., 23
Hales, S., 146
Hall, G.S., 206
Halley, E., 50, 51
Hankwitz, G., *see* Godfrey, A.
Harley, R., 176, 192, 194, 204, 238
Harradence, R., 10
Harriot, T., 34
Harrison, J., 263
Harvard, J., 57, 69, 117
Harvey, W., 3, 19, 145, 163
Hassel, O., 118
Hay, W.T., 54
Heatley, N., 199
Helmholtz, H. von, 231
Henrique, I.D., *see* Henry, P.
Henry VIII, 144, 153, 155, 262
Henry, J., 223
Henry, P., 32
Herschel, C.L., 52, 74, 227
Herschel, F.W., 7, 52
Hertz, H., 226
Hewson, W., 146
Higgs, P., 100
Hill, A.V., 23, 54, 119
Hipparchus, 94

Hippocrates, 140, 143, 147, 163
Hodgkin, D., 227
Hodgkin, T., 196, 210
Hoffenberg, R., 148
Hofmann, A.W. von, 108, 109, 111, 112
Hogg, A., 257
Hogg, Q., 257
Holmes, O.W., 186
Holmes, T., 160
Homer, 79
Hooke, R., 50, 51, 61, 65, 80, 146
Hooker, J.D., 22, 126, 127, 129, 131, 134, 135, 138
Hooker, W.J., 126, 127, 129
Hopkins, F.G., 23, 114, 116
Horsley, V., 213, 214
Horton, J.A., 152
Howard, L., 53
Hudson, W.H., 42, 136, 137
Huggins, M.L., 225
Huggins, W., 134, 225
Hughes, D.E., 226
Humboldt, A. von, 108
Hume, D., 4, 5, 19
Hunter, J., 180, 189, 209
Hunter, W., 19, 146, 180, 209
Hutchinson, J., 211
Huxley, A., 135
Huxley, J., 135, 138
Huxley, L., 138
Huxley, T.H., 22, 127, 131, 134, 135, 138
Hygeia, 141, 161

I
Inglis, E., 173
Ingold, C., 115

J
Jackson, J.H., 203
James I, 34, 145, 248
Jefferies, R., 137
Jefferson, T., 4
Jenner, E., 148, 165, 188, 189, 209
Jenssen, P., 111
Jesty, B., 188, 189
Jesus Christ, 193
Johnson, A., 259, 260
Jones, E., 206
Joule, J.P., 88, 106
Jung, C., 206
Justinian, 19

K
Kant, I., 75
Kármán, T. von, 96
Keats, J., 161–163
Kelvin, L., 21, 88, 118, 180, 225
Kennedy, J.F., 148
Kepler, J., 34, 82, 94
Kertbeny, K.-M., 207
Kingsley, M., 41
Kirkegaard, S., 5
Klein, E.E., 169
Klein (née Reizes), M., 207
Klug, A., 10, 121, 170

Koch, R., 166
Koestler, A., 268
Körösy, D., 26
Korsakov, S.S., 148
Kratzer, N., 262
Kuhn, T., 122

L
La Place, *see* Laplace, P.S.
Lachmann, G., 259
Lagrange, J.-L., 85
Lamarck, J.-B., 53
Lankester, E.R., 134
Laplace, P.S., 18, 85
Laue, M. von, 77
Laveran, C.L.A., 167
Lavoisier, A., 69, 85, 104, 105, 109
Lawrence, J., 187
Leibnitz, G.W. von, 18, 84
Leibniz, *see* Leibnitz, G.W. von
Leishman, W.B., 167
Leonardo, *see* da Vinci, L.
Lewis, H.K., 143
Lewis, T.R., 167
Liebig, J. von, 107–109
Linacre, T., 144
Lincoln, A., 145
Lind, J., 167, 168
Lindemann, F.A., 268
Lindley, J., 128, 129
Lindley, W., 246
Lindley, W.H., 246
Lindsay (née Murray), L., 175
Lindsay, R., 175
Linnaeus, C., *see* Linné, C. von
Linné, C. von, 18, 123
Lister, J., 165, 180, 186, 190, 210
Lister, W., 157
Livingstone, D., 38, 39
Locke, J., 4, 19, 146
Lockyer, N., 111, 134
Lodge, O., 222
Loewi, O., 197
Logue, L., 192
Lonsdale, K., 227
Lovelace (née Byron), A., 74, 231
Lovelock, J.E., 170
Lyell, C., 131, 238
Lysenko, T., 138

M
MacCormac, W., 160
Mackenzie, J., 203
Mackenzie, M., 191
Makins, G., 150
Manby, C., 265
Mann, H., 142
Mann, I., 176, 195
Manson, P., 164, 165, 169
Marconi, G., 222, 226, 228
Markham, C.R., 44, 48
Marsden, W., 211
Marx, K., 15
Mary I, 58
Mason, B., 239

Massey, E., 262
Maxim, H., 267
McCall, A., 173
McCance, R., 117
McGowen, H., 103
McIndoe, A., 213, 214
McLaren, A., 227
Mead, R., 156
Medawar, P., 190, 191
Mendel, G., 90
Mendeleev, D.I., 69, 110, 113
Merrick, J., 213
Meryon, E., 201
Meyer, L.J., 110
Meyerhof, O.F., 23
Michelangelo, 19, 220
Milton, J., 19, 86
Mitchell, R.J., 260, 261
Moissan, H., 69
Mond, A., 103, 240
Mond, L., 103, 240
Moor, J., 262
More, T., 57, 144
Morris, W., *see* Nuffield, L.
Morse, S.F.B., 223
Mortimer, E., *see* Harley, R.
Muirhead, A., 221, 222
Murchison, C., 159, 160
Murrell, C., 174
Murrow, E.R., 229
Myddelton, H., 247, 248

N
Napoleon, 70, 85
Napoleon III, 183
Nasmyth, J., 238
The Navigator, *see* Henry, P.
Nemon, O., 116, 205
Neumann, J. von, 96, 232, 268
Newcomen, T., 252
Newlands, J.A.R., 109, 110
Newton, I., 6, 13, 15, 19, 21, 51, 65, 66, 74, 75, 78, 80, 82–84, 88, 90, 101, 106, 145, 156
Nightingale, F., 153, 155, 160, 171, 178, 183
Nissel, G., 195
Nobel, A., 103, 140, 240
Nuffield, L., 161, 162

O
Oates, L., 45
Oliphant, M., 10
Oliver, P.L., 214, 215
Osler, W., 149
Oursian, N., 262
Owen, R., 129, 134

P
Page, F.H., 258, 259
Palmerston, L., 131
Parkes, D., 262
Parkes, E.A., 167, 168
Parkinson, James, 200, 201
Parkinson, John, 203
Parr, T., 145

Parry, W.E., 43
Pasteur, L., 109, 167, 168, 187, 214
Peacock, G., 231
Pearsall, P., 239
Pearson, K., 90
Pedley, F.N., 162
Peel, R., 109, 183
Penzias, A., 25
Perkin, A.G., 112
Perkin, W.H., 109, 112
Perkin, W.H. Jr., 112
Perrin, J., 125
Pettenkofer, M. von, 168
Pinchbeck, C., 262
Pitt, W., the Elder, 265
Pitt, W., the Younger, 265
Plato, 18, 19
Polanyi, M., 268
Pomerance, B., 213
Popper, K., 4
Porter, G., 78, 118
Preece, W.H., 134
Priestley, J.B., 15, 103, 104
Pringle, J., 168
Pythagoras, 2, 11, 79

R
Rae, J., 42, 43
Rakos, I., 195
Raleigh (Ralegh), W., 34, 35, 49
Ramsay, W., 100, 111, 113, 114
Raphael, *see* Santi, R.
Ray, J., 123, 137
Rayleigh, L., 23, 113, 134, 225
Reed, W., 168
Rees, J.M., 192
Rennie, J., 237, 241, 242
Reuter, P.J., 224
Reynolds, J., 19
Ricardo, H., 256
Richardson, O., 91
Robinson, J., 182
Robinson, R., 112
Rockefeller, J.D., 149
Rolls, C., 256, 259
Ronalds, F., 222
Roosevelt, F.D., 96, 97
Roosevelt, T., 40
Ross, C., 258
Ross, J.C., 42, 127
Ross, R., 168, 169
Rothschild, M., 67
Rothwell, N., 152
Roy, W., 238
Royce, H., 256
Rumford, C., 68–70
Russell, B., 91
Rutherford, E., 21, 23, 77, 93

S
Salam, M.A., 93
Salvin, O., 136
Santi, R., 19
Saunders, C., 155, 156
Saunders, E., 212

Sayers, D.L., 118, 119
Sayre, L.A., 160
Scheele, C.W., 104
Schiller, F., 86
Schlesinger, A. Jr., 145
Schweitzer, A., 148
Scott, R.F., 44, 45
Scott, R.H., 134
Scott, W., 74
Seacole, M., 153, 178, 179, 235
Selous, F.C., 40
Semmelweis, I., 186
Shackleton, E., 44, 46
Shakespeare, W., 5, 6, 75, 145
Shattuck, L., 169
Sherlock, S., 177
Short, E., 259
Short, H., 259
Short, O., 259
Shukhov, V.G., 17
Siebe, A., 265
Simms, J., 262
Simon, J., 159, 160, 165
Simpson, C.K., 196
Simpson, J.Y., 151, 182
Sims, J.M., 131, 150
Sloane, H., 13, 15, 17
Smith, A., 19, 252
Smith, F.P., 256
Smith, J., 35
Smith, Lord of Marlow, 150
Smith, W., 171, 237
Smoluchowski, M., 125
Snow, J., 166, 182–184, 186, 245
Solly, S., 158
Solvay, E., 240
Somerville (née Fairfax), M., 52, 74, 227, 231
Somerville, W., 74
Somorjai, G., 26
Sopwith, T., 259
South, J.F., 158
Speke, J.H., 39, 40
Spilsbury, B., 196
St Thomas, Apostle, 153
Stanhope, C., 265
Stanhope, E., 21, 264
Stanley, H.M., 38, 39
Stephenson, G., 254
Stephenson, M., 227
Stephenson, R., 254, 265
Still, G.F., 203, 204
Stokes, A.R., 120, 121
Stokes, G.G., 134
Stopes, M., 175, 176
Strutt, J.W., *see* Rayleigh, L.
Suess, E., 238, 239
Swift, J., 35
Sydenham, T., 147, 165
Sylvester, J.J., 134
Szilard, L., 96, 268

T
Telesphorus, 141
Teller, E., 96
Theed, W. Jr., 3, 4, 19, 239
Theiler, M., 170

Thompson, B., *see* Rumford, C.
Thomson, G.P., 91, 93
Thomson, J.J., 21, 70
Thomson, W., *see* Kelvin, L.
Titian, 19, 220
Tizard, H.T., 267, 268
Todd, R.B., 194, 202
Tompion, T., 262, 263
Torricelli, E., 94
Toynbee, J., 212
Tréfouël, J., 116
Treves, F., 213
Trevithick, R., 253
Tu, Y., 169
Turing, A.M., 217, 232–235
Turner, E., 101, 107
Turner, J.M.W., 74
Turner, T., 155, 156
Turner-Warwick, M., 177
Tyndall, J., 75, 76, 134, 135

U
Upjohn, J., 262

V
Verdon Roe (Verdon-Roe), E.A., 258
Victoria, 3, 7–12, 24, 28, 37, 47, 75, 88, 131, 182, 212–214
Virgil, 79

W
Wakley, T., 127, 151, 210
Waksman, S.A., 116
Walker, J., 241–243
Walker, J.H., 173
Wallace, A.R., 22, 47, 125, 131, 132, 239
Wallenberg, R., 26
Ward, S., 60
Warington, R., 101
Waterston, D., 203

Waterton, C., 46, 47
Watson, J.D., 17, 119–122
Watson-Watt, R., 261
Watt, J., 2, 180, 251–253, 269
Weizmann, C., 240, 241
Wellcome, H.S., 148, 216
Wells, H.G., 98, 138
Welsch, M., 116
Wernher, J., 236
Wheatstone, C., 222, 223
Whitehead, A.N., 91
Whittle, F., 20
Widdowson, E., 116, 227
Wigner, E.P., 96, 268
Wilberforce, B., 131
Wilkins, A.F., 261
Wilkins, J., 60
Wilkins, M.H.F., 120, 121
Wilkinson, G., 117
Willan, R., 150, 201
William IV, 148
Williamson, A.W., 134
Willis, T., 146
Willughby, F., 123
Wilson, E.A., 44, 45
Wilson, H.R., 120, 121
Winsor, F., 266
Winzer, F., *see* Winsor, F.
Wittgenstein, L., 162, 233
Wöhler, F., 108
Woodall, C., 266, 267
Wren, C., 11, 12, 19, 49–51, 58, 60, 62, 64, 65, 220
Wright Brothers, 256

Y
Yearsley, J., 191
Young, T., 87

Z
Zvorykin, V.K., 229
Zweig, S., 206

Index of Artists and Architects

A
Adams, G.G., 158
Adlard, H., 87
Alexander, W., 104
Allen, J., 106
Amery, S., 155
Anastasi, G., 262
Anrep, B., 24
Armstead, H., 3, 5, 28, 31, 33, 36, 38, 42, 56, 79, 140, 251
Aumonier, E., 141
Aumonier sculptor firm, 143

B
Baayes, G., 103
Bacon, C., 11
Baines, F., 102
Barraud, 167
Bassano Ltd., 167
Beale, M., 60
Behnes, W., 58, 147, 157
Belliard, Z., 107
Benzoni, G.M., 147
Black, M., 224
Blaikley, A., 8
Blake, W., 15, 80, 81
Board, E., 6, 26, 62, 181
Boehm, J., 12, 56–138
Boonham, N.F., 150, 209
Brock, T., 11, 27, 28, 71
Bulman, B., 116
Burne-Jones, P., 113
Burton, I., 39
Butler, J., 249
Butler, T., 107

C
Cartwright, T., 155
Chalon, A.E., 231
Champneys, B., 132
Chantrey, F., 87, 106, 124, 251
Cochin, C.N., 85
Cole, B., 101
Colton, W.R., 40
Cook, H., 144
Cope, A.S., 112
Cornish, M., 67

D
Damer, A.S., 124
Dawson, R., 155
Day & Haghe, 108
Daymound, J. & Sons, 3, 57
Deneulain, 173
Dequevauviller, F.J., 262
Dew-Smith, A.G., 41
Diethe, R., 52
Doubleday, J., 243
Downey, W. & D., 88
Drury, A., 19, 160, 251
Duboy, P., 150
Durham, J., 11, 19, 81, 139

E
Edwards, E., 101
Epstein, J., 24, 93

F
Faber, J., 51
Farmer and Brindley, 2–26
Finch, P., 203
Foggini, B., 80
Foley, J.H., 11, 30, 71
Fontana, G., 5
Ford, O., 135
Frampton, G., 179, 257

G
Gilbert, A., 106
Gillray, J., 69
Glindoni, G., 6
Grant, E., 132
Greenhill, J., 60
Greer, R., 65

H
Hancock, J., 2
Hardiman, A., 200
Holbein, H., 262
Holloway, M., 219
Houston, R., 156
Howard, H., 70

Hudson, T., 146
Hughes, G., 232
Huxley-Jones, T.B., 38
Huysmans, J., 146

J
Jackson, J.H., 20, 203
Jagger, C.S., 46
Jennings, M., 153
Johnson, C., 248
Jones, H., 12, 250
Joseph, S., 247

K
Kent, W., 21
Kerrison, H., 215
Kettle, S., 260
Keyworth, W.D., the Elder, 158
Kindersley, R., 51
Kirby, R., 154
Kneller, G., 62
Kriehuber, J., 168

L
Lacey, W., 9
Lafosse, J.B.A., 109
Lambert, M., 161
Lawrence, T., 87
Lely, P., 60
Levachez, C.F., 105
Lock & Whitfield, 132
Lockyer, J., 108
Lutyens, E., 174

M
MacCarthy, H.P., 151
Maguire, T.H., 125–128, 130
Marochetti, C. (Charles), 243, 254
Marshall, W.C., 81, 189
Maull & Polyblank, 129
Mayall, J.E., 9, 254
McArdell, J., 146
McDovell, P., 123, 126
Mestorovic, I., 173
Monteverde, G., 189
Mountford, E.W., 83
Müller, T., 69
Mullins, E.R., 159
Murray, T., 35, 51

N
Nadar, 167
Nemon, O., 116, 205
Newman and Billing, 163
Nicholl, W.G., 3, 81
Noble, M., 4, 19, 41, 70, 180, 202
Nollekens, J., 60

P
Paolozzi, E., 15, 80, 264, 269
Peale, C.W., 46

Philip, J.B., 2, 3, 5, 28, 33, 36, 38, 42, 56, 140, 251
Phillips, C., 157
Phillips, J., 74
Pierson, P.-L., 109
Plazzotta, E., 219
Pollock, H., 180
Pomeroy, F.W., 3, 251
Poole, D., 177
Pound, D.J., 254
Pugin, A., 21
Pye, C., 62

Q
Quinton, M., 91

R
Ramsay, A., 156
Ravera, J., 241
Reeves, H., 255
Richards, M., 37
Ridley, 124
Riviere, H.G., 215
Rizzello, M., 241
Rosenberg, C., 104
Roubiliac, L.-F., 15, 123
Rowlandson, T., 21
Russell, J., 124
Rysbrack, M., 13, 15, 21, 49

S
Salomon, H.H., 149, 164, 186, 225
Sawyer, 109
Scheemakers, P., 5, 155, 161
Schmutz, R., 102
Schwerdgeburth, C.A., 86
Scott, G.G., 11
Scott, K., 44
Scott, T.D., 230
Scriven, E., 49
Seale, G.W., 57
Seale, J.W., 163
Shannan, A.M., 225
Sharp, M.W., 188
Shawcross, C., 122
Sherwood, J., 18, 19
Simonds, G.B., 244
Sims, 131
Smallwood, W.F., 20
Smirke, R., 14
Smith, A., 131
Smith, J.R., 50, 149
Smith, S., 13
Stephens, E.B., 64, 220
Stevenson, J.A., 46
Sustermans, J., 80
Svarochek, 39

T
Tardieu, A., 166
Theed, F., 159
Theed, W. Jr., 3, 4, 19, 30, 239
Thomas, J., 248

Index of Artists and Architects

Thomson, J., 70
Tookey, J., 263
Turnerelly, P., 251

V
Vanderbank, J., 66
Vertue, G., 102

W
Walery, 111
Walker, A.G., 155, 174, 178, 266
Ward, L., 110, 113

Waterhouse, A., 17
Weekes, H., 144, 156, 157
Westmacott, J.S., 18, 19
Westmacott, R., 4
Williams, J.M., 245
Williamson, S., 161
Wilton, J., 147
Wirgman, T.B., 75
Woodington, W.F., 18, 19, 79
Worthington, W.H., 106
Wren, C., 51, 60, 62, 64, 65
Wright, F.A., 136
Wright, J., 149
Wyon, E.W., 18, 80, 85

GPSR Compliance

The European Union's (EU) General Product Safety Regulation (GPSR) is a set of rules that requires consumer products to be safe and our obligations to ensure this.

If you have any concerns about our products, you can contact us on

ProductSafety@springernature.com

In case Publisher is established outside the EU, the EU authorized representative is:

Springer Nature Customer Service Center GmbH
Europaplatz 3
69115 Heidelberg, Germany

www.ingramcontent.com/pod-product-compliance
Lightning Source LLC
LaVergne TN
LVHW080135260326
834688LV00042B/1178